Collaborative Futures

Table of Contents

The Present

Futures

Epilogue

Appendices

Introduction

1. Anonymous

- *You do not talk about Anonymous.*
- *You do NOT talk about Anonymous. (Wikis are fine though. FEAR US.)*
- *Anonymous works as one, because none of us are as cruel as all of us.*
- *Anonymous is everyone.*
- *Anonymous does it for the lulz.*
- *Anonymous cannot be out-numbered, Anonymous out-numbers you.*
- *Anonymous is a hydra, constantly moving, constantly changing. Remove one head, and nine replace it.*
- *Anonymous reinforces its ranks exponentially at need.*
- *Anonymous has neither leaders nor anyone with any higher stature.*
- *Anonymous has no identity.*
- *Anonymous is Legion.*
- *Anonymous does not forgive.*
- *Anonymous does not forget.*

13 of the 41 entries in the Sekrit Code of Anonymous

In this section we're breaking the first two rules of the Sekrit Code of Anonymous <*encyclopediadramatica.com/index.php? title=Anonymous&oldid=1998440936#The_Sekrit_Code*>. When others have done this in the past it has brought down the wrath of this shadowy group of anonymous individuals, causing public humiliation, hacked servers, and other florid forms of chaos.

Anonymous is a collection of individuals that post anonymously on /b/ <*img.4chan.org/b/*>, a section of the image board 4chan.org. When you post content on a typical message board, you are often required to enter your name. If you don't, your entry is attributed to "anonymous". On /b/ everyone posts as "anonymous". The collective actions of users identified with the name anonymous aggregates into the collective identity Anonymous.

The majority of Anonymous' activity is visible only to Anonymous. The members trade images and jokes between one another on 4chan and other sites. They traffic in pornography, shock imagery, and inane jokes. They collect and distribute the oddities of the web. However, Anonymous is also responsible for occasional external, organized actions—ranging from pranks done "for the lulz", to large scale activist projects. The most visible and longest lived of such projects is called Project Chanology, and is a large scale, distributed war on The Church of Scientology. The first major incident in this war was Anonymous' distribution of a "internal-use only" video featuring Tom Cruise, and Scientology's attempted suppression of the same. Soon after, the declaration of war was made formal, and posted to YouTube (anonymously, of course). Narrated by a text-to-speech generator, the video outlines Anonymous' issues with Scientology:

> "Hello, Scientology. We are Anonymous.
>
> Over the years, we have been watching you. Your campaigns of misinformation; suppression of dissent; your litigious nature, all of these things have caught our eye. With the leakage of your latest propaganda video into mainstream circulation, the extent of your malign influence over those who trust you, who call you leader, has been made clear to us. Anonymous has therefore decided that your organization should be destroyed. For the good of your followers, for the good of mankind—for the laughs—we shall expel you from the Internet and systematically dismantle the Church of Scientology in its present form. We acknowledge you as a serious opponent, and we are prepared for a long, long campaign. You will not prevail forever against the angry masses of the body politic. Your methods, hypocrisy, and the artlessness of your organization have sounded its death knell."
> <*www.youtube.com/watch?v=JCbKv9yiLiQ*>

Since then, Anonymous has mounted repeated electronic attacks on Scientology websites, coupled with large scale protests outside of Scientology centers across the world. Throughout this large scale, coordinated, goal oriented collective action no one has emerged as the leader to speak for the group. In fact, no one has spoken to the press at all, though the press has reported extensively on the events. The only communiques come in the form of anonymously posted videos and anonymous posts to /b/ with instructions for when to protest, how to conduct yourself during the protests, what to wear, etc.

In this book we attempt to articulate what constitutes a collaboration. We argue that rules for participation, established guidelines for attribution, organizational structure and leadership, and clear goals are necessary for collaboration. In most cases, when we think of these attributes, we think of manifestos of artist and activist groups, attempts to govern attribution by formal licenses like the Free Culture and Free Software licenses, Debian's formal decision making process, or Eric Raymond's notion of a Benevolent Dictator that characterizes Linus Torvald's governance over Linux.

What is fascinating about Anonymous, is that at first glance, it appears they have none of these: They are often portrayed as a band of predominantly young white male renegade hackers raining chaos on random corners of the Internet with no logic or reason. They have even been called Terrorists. But in fact, their *Sekrit Code* establishes clear rules. Participation requires posting as Anonymous and not talking about Anonymous. Attribution is strictly collective and anonymous under a unified group identity. The organizational structure is clear: There are no "leaders nor anyone with any higher stature." The code even establishes goals: "the lulz" adapted from "LOL", in short, for kicks.

Anonymous has operated under rules that are directly opposed to the rules that have governed most successful large-scale collaborations. How then do goals as broadly defined as "the lulz" become defined and articulated into a goal like the intent to "systematically dismantle the Church of Scientology"? How can an organization with no leaders articulate and execute such an ambitious and "long, long campaign"? How can the enforced absence of any structure as a governing principle result in such effective and coordinated action?

Is this a possible collaborative future? If so, it is a terrifying one in which anonymity and structurelessness permits total absolution of social responsibility, terrorizing of innocent outsiders, and harassment of those who provide public feedback, criticism and indeed even speak of the group ("You do not talk about anonymous"). It is a P2P, collaborative, digitized "Lord of the Flies" wherein boys' games devolve into violence for fun. In the perpetual techno-utopian dialectic, this is the feared dystopian future we hope will be avoided, as we aim for the utopia that we can never actually arrive at.

2. How this Book is Written

"Collaboration on a book is the ultimate unnatural act."
—Tom Clancy

This book was first written over 5 days (Jan 18-22, 2010) during a Book Sprint in Berlin. 7 people (5 writers, 1 programmer and 1 facilitator) gathered to collaborate and produce a book in 5 days with *no prior preparation* and with the only guiding light being the title 'Collaborative Futures'.

These collaborators were: Mushon Zer-Aviv, Michael Mandiberg, Mike Linksvayer, Marta Peirano, Alan Toner, Aleksandar Erkalovic (programmer) and Adam Hyde (facilitator).

The event was part of the 2010 transmediale festival <*www.transmediale.de/en/collaborative-futures*>. 200 copies were printed the same week through a local print on demand service and distributed at the festival in Berlin. 100 copies were printed in New York later that month.

This book was revised, partially rewritten, and added to over three days in June 2010 during a second book sprint in New York, NY, at the Eyebeam Center for Art & Technology as part of the show *Re:Group Beyond Models of Consensus* and presented in conjunction with Not An Alternative and Upgrade NYC.

In the execution of code, the computer could care less about different styles of code, which design patterns are used, and where the curly brackets go. But in the human world, inconsistency becomes information. While stylistically different code can be flattened to a singular, executable voice, inconsistent human communication is harder to process, "decode", and unravel (See the Can Design By Committee Work chapter).

In this book there are inconsistencies, occurring in the shifts between the distinct voices that constitute the text in its entirety. By constantly re-writing, over-writing, and un-writing the book, the residual material (that which remains unseen in the printed version) is also the material that expresses the mode of collaboration at work here. Each collaborative (futures) book is fundamentally a reference to a very particular micro-community. In this sense it can be seen as attributing to a social study. There is no generality in collaboration.

Three new core members joined Mushon Zer-Aviv for the duration of the project in New York: Astra Taylor, kanarinka, and sissu. Michael Mandiberg, Mike Linksvayer, Alan Toner, Adam Hyde and Marta Peirano joined at various times in person and online.

A brief outline of the calendar, methodology and participants can be found in the appendices "Anatomy of the First Sprint" and "Anatomy of the Second Sprint".

What is a Book Sprint?

"A book is a place where readers and writers meet"
—Bob Stein, Institute for the Future of the Book

The Book Sprint concept was devised by Tomas Krag. Tomas conceived of book production as a collaborative activity involving substantial donations of volunteer time.

Tomas pioneered the development of the Book Sprint as a 4 month+ production cycle, while Adam Hyde, founder of FLOSS Manuals, was keen to continue with the idea of an "extreme book sprint," which compressed the authoring and production of a print-ready book into a week-long process.

During the first year of the Book Sprint concept FLOSS Manuals experimented with several models of sprint. So far about 16 books have been produced by FLOSS Manuals sprints, some of these were 5 day sprints, but there have also been very successful 2 and 3 day events.

Because Book Sprints involve open contributions (people can contribute remotely as well as by joining the sprint physically) the process is ideally matched to open/free content. Indeed, the goal of FLOSS Manuals embodies this freedom in a two-fold manner: it makes the resulting books free online, and focuses its efforts on free software.

FLOSS Manuals has produced many fantastic manuals in 2-5 day Book Sprints. The quality of these books is exceptional, for example Free Software Foundation Board Member Benjamin Mako Hill said of the 280 page Introduction to the Command Line manual (produced in a two day Book Sprint):

"I have written basic introductions to the command line in three different technical books on GNU/Linux and read dozens of others. FLOSS Manual's "Introduction to the Command Line" is at least as clear, complete, and accurate as any I've read or written. But while there are countless correct reference works on the subject, FLOSS's book speaks to an audience of absolute beginners more effectively, and is ultimately more useful, than any other I have seen."

But Collaborative Futures is markedly different. The difference between the Collaborative Futures and other Book Sprints is that this is the first sprint to make a marked deviation from creating books which are primarily procedural documentation. To ask 5 people who don't know each other to come to Berlin and write a *speculative narrative* in 5 days when all they have is the title is a scary proposition. To clearly define the challenge we did no discussion before everyone entered the room on day 1. Nothing discussed over email, no background reading. Nothing.

Would we succeed? It was hard to consider this question because it was hard to know what might constitute success. What constituted failure was clearer —if those involved thought it was a waste of time at the end of the 5 days this would be clear failure. All involved had discussed with the facilitator the possibility that the project might fail (transmediale also discussed this with the facilitator).

Additionally, as if this was not hard enough, we decided to use the *alpha* version of a new collaborative platform 'Booki' <*www.booki.cc*>. One of the Booki developers (there are two)—Aleksandar Erkalovic—joined the team in Berlin to bug fix and extend the platform as we wrote.

We also had to develop new methodologies for this sprint. Try new things out, test ideas and review their effectiveness. All in 5 days.

As a result we have a book, a vastly improved (free) software platform, happy participants, and clear ideas on what new methods worked and what didn't. We look forward to your thoughts and contributions… See **Write this Book** in the **Epilogue**.

 Glossary: **Glossary (unconscious/semiconscious/conscious)**

Glossary items are distributed across the book according to their appropriate place—relating to particular themes. The specific format gives them a distinctive voice to differ from the main body of text.

The glossary is a way of elaborating on a number of terms and expressions that, to some degree, form the kernel of the book. A focus is given to the semi-conscious and un-conscious dimension of this glossary (of any glossary?!). While some terms are clearly major threads running through the discussion and throughout the book, others pop up intuitively. The glossary is always also fiction, and supplementary.

:

Art++ | Architecture | Autonomy | Bike-Shedding | Collaboration |
Coming | Contract = temporary contract (friendship and otherness)
| Discourse | Dissent | Distribution | Educational Intervention |
Extraction | Free | Google Wave | Inconsistencies | Imaginary
Reader | Invitation | Location-Locating | Minor | Mythologies | Non-
Documents | Non-human Collaboration | Open | Progress |
Acceleration/Deceleration | Speed | The Glossary of Tyranny |
Tyranny | Vocabularies

Thanks

Many thanks to Stephen Kovats who supported this enterprise with
conviction. Without Stephen's commitment to the project it would not have
been possible.

Thanks to the curators of the Re:Group show, Eyebeam, Not An Alternative
and Upgrade New York for hosting and supporting the second edition book
sprint.

Also thanks to Laleh Torabi for designing the first cover and to Galia Offri
for designing the second cover.

Collaborative Futures - First Edition cover by Laleh Torabi

Collaborative Futures - Second Edition cover by Galia Offri

3. A Brief History of Collaboration

> Whenever a communication medium lowers the costs of solving collective action dilemmas, it becomes possible for more people to pool resources. And "more people pooling resources in new ways" is the history of civilization in...seven words.
>
> —Marc Smith, Research sociologist at Microsoft

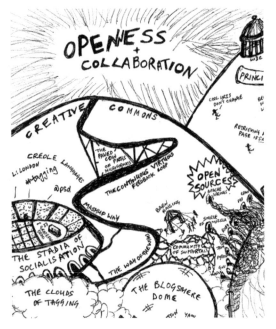

Detail of The Web is Agreement / Paul Downey / CC BY

This book is about the future of collaboration; to get there, it is necessary to understand collaboration's roots.

It is impossible to give a full history in the context of this book; we instead want to highlight a few key events in the development of collaboration that directly inform the examples we will be looking at.

Most of these stories are well known, so we decided to keep them short. They are all very well documented, so these descriptions should be great starting points for further research.

Anarchism in the Collaboratory

Anarchist theory provides some of the background for our framing of autonomy and self organization.

This is recapitulated by Yochai Benkler, one of the leading modern theorists of open collaboration, in his book The Wealth of Networks: How Social Production Transforms Markets and Freedom:

> "The networked information economy improves the practical capacities of individuals along three dimensions: (1) it improves their capacity to do more for and by themselves; (2) it enhances their capacity to do more in loose commonality with others, without being constrained to organize their relationship through a price system or in traditional hierarchical models of social and economic organization; and (3) it improves the capacity of individuals to do more in formal organizations that operate outside the market sphere. This enhanced autonomy is at the core of all the other improvements I describe. Individuals are using their newly expanded practical freedom to act and cooperate with others in ways that improve the practiced experience of democracy, justice and development, a critical culture, and community.
>
> ...
>
> [M]y approach heavily emphasizes individual action in nonmarket relations. Much of the discussion revolves around the choice between markets and nonmarket social behavior. In much of it, the state plays no role, or is perceived as playing a primarily negative role, in a way that is alien to the progressive branches of liberal political thought. In this, it seems more of a libertarian or an anarchistic thesis than a liberal one. I do not completely discount the state, as I will explain. But I do suggest that what is special about our moment is the rising efficacy of individuals and loose, nonmarket affiliations as agents of political economy."

 Glossary: **Non-human Collaboration**

Why do we imagine it is only humans who act, react and enact the world? What if plants, animals, things, forces and systems can also exert agency? Who are we collaborating with and through? Living things and inert matter. Organisms of all kinds could be included in forms and assemblages of collaboration. The agents: human and non-human entities, plants, objects, systems, histories. This thinking proposes an inclusive model.

Science to Software

Although the history of science is intertwined with that of states, religions, commerce, institutions, indeed the rest of human history, it is on a grand scale the canonical example of an open collaborative project, always struggling for self-organization and autonomy against pressure from state, religion, and market, in a quest for a common goal: to discover the truth. Collaboration in science also occurs at all timescales and levels of coupling, from deeply close and intentional collaboration between labs to opportunistic collaboration across generations as well as problematic collaborations been researchers and industry.

Glossary: **Progress**

> Understood as a belief system. An inheritance of the Enlightenment akin to the idea of a singular, objective "truth" out there awaiting human discovery. Belief in the perpetual improvement of things (an easier way of living and acting, or society as such) via the development of technology. Progress as the fetishism of change and constant transformation. See also Speed and Acceleration/Deceleration.

The last half millennium produced innumerable examples of interesting collaboration in addition to the great scientific endeavor. Within the technological sphere, none is as cogent in informing and driving contemporary collaboration as the Free Software movement, which provides much of the nuts and bolts immediate precedent for the kinds of collaborations we are talking about—and often provides the virtual nuts and bolts of these collaborations! The story goes something like this: Once upon a time all software was open source. Users were sent the code, and the compiled version, or sometimes had to compile the code themselves to run on their own specific machine. In 1980 MIT researcher Richard Stallman was trying out one of the first laser printers, and decided that because it took so long to print, he would modify the printer driver so that it sent a notice to the user when their print job was finished. Except this software only came in its compiled version, without source code. Stallman got upset—Xerox would not let him have the source code. He founded the GNU project and in 1985 published the GNU Manifesto. One of GNU's most creative contributions to this movement was a legal license for free software called the GNU Public License or GPL. Software licensed with the GPL is required to maintain that license in all future incarnations; this means that code that starts out freely licensed has to stay freely licensed. You cannot close the source code. This is known as a Copyleft license.

Mass Collaborations

Debian is the largest non-market collaboration to emerge from the free software movement.

Beginning in 1993, thousands of volunteer developers have maintained a GNU/Linux operating system distribution, which has been deeply influential well beyond its substantial deployments.

Debian has served as the basis for numerous other distributions, including the most popular for the past several years, Ubuntu. Debian is also where many of the pragmatics of the free software movement were concretized, including in the Debian Free Software Guidelines in 1997, which served as the basis of the Open Source Definition in 1998.

In 1995 Ward Cunningham created the first wiki, a piece of software that allowed multiple authors to collaboratively author documents. This software was used especially to hold meta-discussions of collaboration, in particular on MeatballWiki. Dozens of wiki systems have been developed, some with general collaboration in mind, others with specific support for domain-specific collaboration, for example Trac for supporting software development. In 2001 Wikipedia was founded, eventually becoming by far the most prominent example of massive collaboration.

Web 2.0 is bullshit

Although they have a fairly distinct heritage, Wikipedia and wikis in general are often grouped with many later sites under the marketing rubric "Web 2.0".

While many of these sites have ubiquitous "social" features and in some cases are very interesting collaboration platforms, particularly when considering their scale, all have extensive precedents.

The Web 2.0 term is directly borrowed from software release terminology.

It implies a major "dot release" of the web—an all encompassing new version, headed by the proprietary new media elite (the likes of Google, NewsCorp, Yahoo, Amazon) that passive web users, still using the old "1.0 version", should all upgrade to. "Web 2.0" also gave birth to the use of "Web 1.0" which stands for conservative approaches to using the web that are merely attempting to replicate old offline publishing models.

More than anything else this division of versions implies a shift in IT business world—an understanding that a lot of money can be made from web platforms based on user production.

This new found excitement of the business sector has brought a lot of attention to these platforms and indeed produced some excellent tools.

But the often too celebratory PR language of these platforms has affected their functionality, reducing our social life and our peer production to politically correct corporate advertising. Sharing, friendship, following, liking, poking, democratizing... collaborating.

These new platforms use a pleasant social terminology in an attempt to attract more users.

But this polite palette of social interactions misses some of the key features that the pioneering systems were not afraid to use.

For example, while most social networks only support binary relationships, Slashcode (the software that runs Slashdot.org, a pioneer of many features wrongly credited to "Web 2.0") included a relationship model that defined friends, enemies, enemies-of-friends, etc.

The reputation system on the Advogato publishing tool supported a fairly sophisticated trust metric, while most of the more contemporary blog platforms support none.

Web 3.0 is also bullshit

"The future is already here—it is just unevenly distributed."
—William Gibson

One might argue that Web 2.0 has popularized collaborative tools that have earlier been accessible only to a limited group of geeks. It is a valid point to make.

Yet the early social platforms like IRC channels, Usenet and e-mail have been protocol based and were not owned by a single proprietor.

Almost all of the current so called Web 2.0 platforms have been built on a centralized control model, locking their users into dependence on a commercial tool.

We do see a turn against this lock-in syndrome.

The past year has seen a shift in attention towards open standards, interoperability and decentralized network architectures.

Glossary: **Google Wave**

In May 2009, Google announced the Google Wave project as its answer to the growing need for more user autonomy in online services and to further its own data-omnivorous agenda. Wave was declared an open source platform for web services that (like email) works in a federated model. What this means is that the user can choose to use a Wave service (on either a Google or a non-Google owned server) through which to channel her data and interaction in an online service.

For example, Alice is using her Google-hosted Wave account to schedule an armed bank robbery with Bob. She is using a Wave-enabled collaborative scheduling application which, in Wave terms, is called a 'Robot'. Bob is using his own encrypted Wave server hosted somewhere in a secret location. Alice is happy since she didn't have to work hard to get the wonderful collaborative functionality of Wave from Google. Bob is happy since he doesn't have to store all his data with Google and can still communicate with his co-conspirator Alice and feel like his data is safe. Google is happy since the entire conversation between the two outlaws is available for it to index and it could have easily targeted them with ads for weapons, ski masks, drilling equipment and vans. In the end, Wave or something like it could be another way in which Open Source software and open standards makes for happy collaboration.

The announcement of Google Wave was probably the most ambitious vision for a decentralized collaborative protocol coming from Silicon Valley. It was launched with the same celebratory terminology propagated by the self-proclaimed social media gurus, only to be terminated a year later when the vision could not live up to the hype.

Web 3.0 is also bullshit. The term was initially used to describe a web enhanced by Semantic Web technologies. However, these technologies have been developed painstakingly over essentially the entire history of the web and deployed increasingly in the latter part of the last decade.

Many Open Source projects reject the arbitrary and counter-productive terminology of "dot releases" the difference between the 2.9 release and the 3.0 release should not necessarily be more substantial than the one between 2.8 and 2.9.

In the case of the whole web we just want to remind Silicon Valley: "Hey, we're not running your 'Web' software. Maybe it's time for you to upgrade!"

Free Culture and Beyond

The Free Culture movement and Creative Commons are built on top of the assumption that there is a deep analogy between writing code and various art forms. This assertion is up for debate and highly contested by some collaborators on this project. (For more on this topic see the chapter "Can Design By Committee Work?")

No doubt software is a cultural form, but we should be aware of the limits of the comparison to other creative modes. After all, software operates according to various objective standards. Successful software works; clean code is preferable; good code executes. What does it mean for a cultural work to "execute"? Where code executes, art expresses. Indeed, many forms of art depend on ambiguity, layered meanings, and contradiction. Code is a binary language, whereas the words used to write this book, even though they are in English, will be interpreted in various unpredictable ways. Looking at all of creativity through the lens of code is reductive.

We also wonder if collaboration is possible or desirable for a project that is deeply personal or subjective. Would I necessarily want to collaborate on a memoir, a poem, a painting? We also wonder if we can ever not collaborate, in the sense that we are always in relationship to our culture and environment, to the creations that proceeded ours, to the audience. Or, to make matters stranger, can we ever not collaborate, even when it seems that we are alone? As musician David Byrne wrote on his blog:

> "But one might also ask: Is writing ever NOT collaboration? Doesn't one collaborate with oneself, in a sense? Don't we access different aspects of ourselves, different characters and attitudes and then, when they've had their say, switch hats and take a more distanced and critical view—editing and structuring our other half's outpourings? Isn't the end product sort of the result of two sides collaborating? Surely I'm not the only one who does this? "

> —David Byrne <*journal.davidbyrne.com/2010/03/031510-collaborations.html*>

For those who believe that the code and culture analogy is deeply insightful, the free culture movement is an attempt to translate the ethics and practices of free software to other fields, some closely tied to technology changes (including wikis and social media sites mentioned above) allowing more access and capability to share and remix materials. Creative Commons, founded in 2001, provides public licenses for content akin to free software licenses, including a copyleft license roughly similar to the FDL (Free Documentation License) that is used by Wikipedia. These licenses have been used for blogs, wikis, videos, music, textbooks, and more, and have provided the legal basis for collaborations often involving large institutions, for example publishing and re-use of Open Educational Resources, most famously the OpenCourseWare project started at MIT as well as many-to-many sharing with extensive latent collaboration, often hosted on sites like Flickr.

 *Glossary: **Art++***

> Aesthetic production can form a coalition with open source and networked culture, with real life as lived outside the art gallery/space/system, and with political concerns such as the Commons, social justice and sustainability. Let us not discount the productive alliances that can be forged amongst areas of experimental cultural practice.

There is still much to learn from historical examples of collaborative theory and practice—and some of these in turn have lessons to learn from current collaboration practices—for anarchist theory, see the Solidarity chapter, for science, see the Science 2.0 chapter. Even the term autonomy may have a useful contribution to contemporary discussion of collaboration, for example resolving the incompleteness and vagueness present in both "free" and "open" terminology.

There are different levels of openness. Being more or less open implies a level of agency, being more or less able to act, and/or being part of something (e.g. a group, debate, project), and/or having the power of access or not. There are logics of control and networks in any collaborative system—but it becomes important to imagine control other than relating to a totality. In their book "The Exploit: A Theory of Networks"(2007), Alexander R. Galloway & Eugene Thacker suggest imagining networks outside an "abstract whole", networks that are not controlled in a total way. They further argue that open source fetishizes all the wrong things. The opposition between open and closed is flawed. An open collaboration in comparison to a less open (almost closed) collaboration suggests the possibility for shifting hierarchies within the collaboration. A closed collaboration can be understood as a defined micro-community that has gathered for particular reasons, and that remains as group intact for the duration of the project.

Collaboration as in Collaborative Futures might be in-between open and closed. The question of a collaborative future is a projection. Using imagination as a tool, a collaborative future is as open as possible confirming the variability of a system - dissent, multiplicity, and possible failure allow agency in its proper sense.

4. This Book Might Be Useless

At the outset we must admit that this book might possibly be useless. Because collaboration is everywhere. To imagine that we could write a book about collaboration is to imagine there is such a thing as *not* collaborating. And to imagine a long history of *not* collaborating with each other. And the ability as "individuals" (Western liberal subjects) to operate separately from the "others" and the world, environment, context around them. And all of that is false! How can we even begin?

Can we be expansive and still say something useful about collaboration? Let us start with breastfeeding.

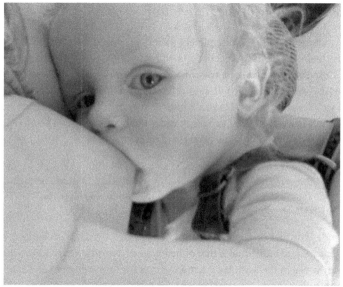

kanarinka's son, March 2009.

Breastfeeding

I always had a sneaking suspicion that I wasn't quite as much myself as I thought I was. It was breastfeeding my son that convinced me of this as a real, material fact. It is very liberating to realize that I am really, wholly not me, that I do not have to figure out "who I am" nor "express myself". My experience of pregnancy and breastfeeding was myself as more than me; not doubled, not serving as a "carrier" for another individual human self. Rather as a joined creature, a multiplication of my creatureliness.

What if we are actually many creatures, many joinings, in many contexts? Not a singular id, ego, and super-ego. Not an integrated self. Not always human. But wholly new creaturely configurations with every step we take, every machine we use, every body part we move, every inhaled breath that alters our body chemistry and exhaled breath that alters the environment.

What if being relational—our relationality—is our primary and sole manner of being and operating in the world?

Why do we think we are so separate from the world to begin with? Why do we think our separate selves would then come together under the rubric of collaborating? Why do we imagine that collaboration might only be possible amongst humans?

A Short History of the Individual

Foucault details how the disciplinary society produced the Western individual—the liberal subject imagined to precede social formations. The individual is imagined to be transhistorical and universal, a basic social "module" which can be combined with other modules (in "collaborations" and various socio-political entities) but not reduced. These individuals are extremely convenient for capitalism. Individuals are framed as having individual desires, individual needs, and individual wishes.

> "We hold these Truths to be self-evident, that all Men are created equal, that they are endowed by their Creator with certain unalienable Rights, that among these are Life, Liberty and the pursuit of Happiness."
> —US Declaration of Independence

So one of the first problems, of course, for the universal, transhistorical, self-evidently equal individual is that of being an exclusionary individual who is not universal in the least. He is a man who is white and straight and middle to upper class. And he is entitled with his inalienable right to pursue his individual happiness which, as we acknowledge in our declaration of individual independence, might potentially be at odds with any kind of communal or public well-being.

This is the perfect ground upon which to cultivate consumer capitalism—an unequaled force of celebration of individual desires, individual needs and individual selves.

Besides, based on at least the past 300 or so years of history since the invention of the individual combined with the relentless reinforcement of capitalism, this is why we imagine we need to write a whole book about when the universal, transhistorical, self-evidently equal individual decides to collaborate. But what if it is just not true? What if individuals—as separate beings-in-themselves—don't even exist?

Glossary: **Collaboration**

A loaded term implicitly linked to formations and formulations of communities, of people together within a work/labour environment. Collaboration has become a strategy and/or style in art, culture and networked structures. There is an assumption that collaboration (in the sense of being more than one making something, more than one working on something) is the preferred working method in order to be properly, truly political and more socially engaged. However, it has been noted, for example, by Maria Lind and Brian Holmes, that there is no non-collaboration in art/culture as such. Rather than generalizing about collaboration, the more salient question would be to singularize collaborative projects and formations, and make clear their specific place, context and potential force in the cultural-political sphere. In parallel, one can then be more explicit about the particular politics at play there. Adopting a kind of radical specificity expands "collaboration" into recurring and urgent questions of the local, the localized, the multicultural, and the side effects, and in return opens out to further analysis, discourse and action.

To summarize: Is this book useless since we are always already collaborating with ourselves, with each other, with our bacteria and public transportation, with our egg-and-cheese sandwiches?

Background Concepts

5. Assumptions

"Xerography—every man's brainpicker—heralds the times of instant publishing. Anybody can now become both author and publisher. Take any books on any subject and custom-make your own book by simply xeroxing a chapter from this one, a chapter from that one—instant steal!

As new technologies come into play, people become less and less convinced of the importance of self expression. Teamwork succeeds private effort. "

—Marshall McLuhan, *The Medium is the MESSAGE*

This book was written in a collaborative Book Sprint by six core authors over a five-day period in January 2010. The six starting authors each come from different perspectives, as are the contributors who were adding to this living body of text.

Six months later a new group of collaborators convened in New York City, while several of the first group also contributed simultaneously from NYC, Berlin and San Francisco. For the most part it can be said that the second sprint brought a fresh set of eyes and a critical perspective to the material produced by the previous group.

Glossary: **Coming**

Coming together. Coming politics! Let us try to proceed with the book as thought—and continue to have a living body of text and follow a certain politics to come.

"Intellectuality and thought are not a form of life among others in which life and social production articulate themselves, but they are rather the unitary power that constitutes the multiple forms of life as form-of-life. In the face of state sovereignty, which can affirm itself only by separating in every context naked life from its form, they are the power that incessantly reunites life to its form or prevents it from being dissociated from its form. The act of distinguishing between the mere, massive inscription of social knowledge into the productive processes (an inscription that characterizes the contemporary phase of capitalism, the society of the spectacle) and intellectuality as antagonistic power and form-of-life—such an act passes through the experience of this cohesion and this inseparability. Thought is form-of-life, life that cannot be segregated from its form; and anywhere the intimacy of this inseparable life appears, in the materiality of corporeal processes and of habitual ways of life no less than in theory, there and only there is thought. And it is this thought, this form-of-life, that, abandoning naked life to 'Man' and to the 'Citizen,' who clothe it temporarily and represent it with their 'rights,' must become the guiding concept and the unitary center of the coming politics."

—Giorgio Agamben, *Means without End: Notes on Politics*. [University of Minnesota Press, Minneapolis, London: 2000]

What this book is...

To begin looking at those futures, we look back to others who have looked into the future. Marshall McLuhan's quote above, from "The Medium is the MESSAGE" give us our first clue about all of these assumptions we are making. We are talking about media, we are talking about freedom, we are talking about technologies, and we are talking about culture.

McLuhan's prophetic utterance, several decades before the photocopier fueled the punk cut-up design aesthetic, or the profusion of home-brew zines, is still unmet. We are still chasing it. Mainstream culture continues to consolidate around block buster films, books, and music. Copyright restrictions make it harder and harder to exercise the creative power of these reproduction tools without breaking increasingly restrictive intellectual property laws.

But one thing is unanimously true: "Teamwork succeeds private effort."

Human beings have always collaborated; collaboration is not a recent development, nor is it rare. The key assumptions we are making in this text are that we are talking about new technologies, that technology is not necessarily computers, that digital media makes it easier to collaborate across distance. We are focused on collaboration that shares similar progressive social goals, and collaboration that is 'free' or 'open' rather than hierarchical production models. We also see a potential threshold between teamwork and collaboration, and between sharing and collaboration.

We are interested in new forms of social organization through online networks. We are excited by the possibility of digital technology to bridge distances: we had collaborators writing this book with us from many corners of the world.

The proliferation of communication networks allows this, as does the invention of new tools for collaboration, but we are also quick to assert that the removal of distance makes other barriers more apparent.

What this book is not...

Despite the fact that the term 'collaborative' has been a buzzword in the art world in recent years, we dedicated little time to it.

Given the complex history of collectivist movements, and the web of relationships present in artists studios and workshops, this was probably advisable.

Collaboration also lies at the heart of the firm, but given the dominance of money in determining participation and the involuntary aspect of work, this aspect is often neglected.

Today the language of 'communities of practice', organization in 'teams', 'self-organised clusters' is ubiquitous in the corporate sphere, as are attempts to enable or capitalize on end-user participation in the production cycle. But this book is not about that.

Finally, collective political movements formed a key force of the twentieth century, and embodied vital instantiations of collaboration. What is to be learned from that history, and how movements are adapting to, or challenged by, the new techniques and organizational forms, represents a vast domain of research beyond the reach of what follows.

This book is not finished.

6. On The Invitation

A shadow hangs over every discussion of social life today: the shadow of economism. In its darkness, light emanates only from productivity, efficiency, incentives and profits. To the epithet of 'dismal' famously ascribed to economics, on account of its pessimistic commitment to the inevitability of scarcity for the many, one could also add 'dark'. The dimness of the prospect derives from the poverty of its instruments of measurement, or rather its reliance on the measurable as a sole index of the 'good'. But the edifice of economic organization sits atop a substrata of generalized cooperation that conditions the very possibility of its existence.

> "The virtually preponderant importance henceforth recognized of externalities bears witness to the limits of the market economy and puts its categories in crisis. It makes visible that the primary sources of wealth, virtual source and condition for all the others, are not manufacturable by any business, accountable in any form of currency, exchangeable against any equivalent. It reveals that the visible economy, so-called formal, is just a relatively small part of the total economy. Its domination over the latter has rendered invisible the existence of a primary economy made up of non-market activities, exchanges and relations through which meaning is produced, the capacity to love, to cooperate, to feel, to link oneself to others, to live in peace with one's body and nature.
>
> It is in this other economy that individuals produce themselves as humans, at the same time mutually and individually, and produce a common culture. The recognition of the primacy of external wealth to the economic system implicates the necessity for an inversion of the relationship between between the production of market "value" and the production of wealth "unexchangeable, unappropriable. intangible, indivisible and unconsumable": the former should be subordinated to the latter.
>
> —Andre Gorz, *The Immaterial, p.80 (translated by Alan Toner)*

Individuals can only produce in connectedness with society, and they do so in manifold ways rarely acknowledged as 'productive'. And it is their horizontal interrelation that enables everything else to function; social production is care outside of crèche, clinic, hospice and hospital; the force for social peace infinitely more powerful than police on the street; the space of educational and cultural development outside of the University and school; the source of skills in language and communication. It is produced every time we ask for directions in the street, recommend a book, help someone carry a pram in the subway, advise on how to fix a pipe, popularize a fashion or lifestyle...

To describe these things as 'productive' seems absurd or even repugnant, because we know and expect that people do not always behave instrumentally. But the temptation to reframe them stems from frustration at the way in which these things are otherwise magically discounted as 'externalities' in a society where economic frameworks and categories are treated as paramount, to the point of rendering anything beyond productivity and profit invisible.

From this perspective, the attention now given to collaboration poses both an opportunity and a problem. The positive aspect is that it represents the emergence from invisibility of that social wealth which comes from beyond market or state, liberating activity from subsumption to market logic. The danger resides in the threat that its recognition *as* productivity may lead to a further reduction of life to economistic categories.

Current hype around collaboration tends to discuss only that part of social production understood as having direct economic impact (because it has been seen to produce substitutes for goods and services previously generated only within the market or the firm). But this aspect is just the tip of the iceberg: the critical project is to bring the rest of the ice into view.

Come in Everybody?

Glossary: **Invitation**

Specific methods by which a project invites the outside in. Linked to questions of having/getting access, entries, hospitality, social bonds, fake bonds.

Like a conversation, creating with others can only take place within a context of willingness: to listen, share, exchange, be contradicted, learn. But what determines how, where and with whom this happens?

Approach a stranger on the street: instead of asking them for directions, ask them if they'll come to the cinema, discuss the film afterwards and jointly compose a review. It is likely that the request will be interpreted as an intrusion, even an aggression, and be rebuffed.

Access to people, spaces and activities is limited by obstacles of class, gender, cultural capital, place and acquaintance. Protocols of participation evolve based on assumptions and norms which are naturalized rather than problematicised. Whilst these modes remain unchallenged their unexamined assumptions are reproduced.

This vantage point allows us to reconsider the nature of the self-selective quality of online participation. When an expanded range of potential contexts is enabled through the attenuation of place as a condition to engagement, other barriers to participation become more visible: you may want to throw yourself into a collective project, but maybe they don't want you. In some cases this may be a matter of pragmatic requirements: thresholds of commitment needed for trust; expectations of responsibility; safety considerations (as in the case of pirate communities or Debian). But there are also exclusions resulting from social stratification or plain cliquishness.

Alternatively invitations can be an expression of a group's willingness to challenge its own composition and practices, making itself available for change.

The invitation is thus an important tool for reproducing and transforming the community. Existing members function as portals through which others can be introduced and vouched for.

But who can accept the invitation? Only those who have spare time, energy and resources, and this is the key limiting factor to any putative hope for 'open participation'. Each participant is circumscribed by their economic and cultural resources. Those who labour at the margins just to survive have little latitude, or energy, to engage in activities geared to enriching their lives in non-material ways.

On Economism and Incentives

If the above addresses the 'with whom' and 'where' can we collaborate, we can now return to the why. But it is an inquiry inflected by the initial comments on social production above. We are always producing with others, and why we do so has varied explanations not all of which can be explained in the language of incentives. Eben Moglen captures this well in his discussion of the energy behind creativity in general and free software in particular:

> "According to the econodwarf's vision, each human being is an individual possessing "incentives," which can be retrospectively unearthed by imagining the state of the bank account at various times. So in this instance the econodwarf feels compelled to object that without the rules [copyright] I am lampooning, there would be no incentive to create the things the rules treat as property: without the ability to exclude others from music there would be no music, because no one could be sure of getting paid for creating it.
>
> ...

The dwarf's basic problem is that "incentives" is merely a metaphor, and as a metaphor to describe human creative activity it's pretty crummy. I have said this before... but the better metaphor arose on the day Michael Faraday first noticed what happened when he wrapped a coil of wire around a magnet and spun the magnet. Current flows in such a wire, but we don't ask what the incentive is for the electrons to leave home. We say that the current results from an emergent property of the system, which we call induction. The question we ask is "what's the resistance of the wire?" So Moglen's Metaphorical Corollary to Faraday's Law says that if you wrap the Internet around every person on the planet and spin the planet, software flows in the network. It's an emergent property of connected human minds that they create things for one another's pleasure and to conquer their uneasy sense of being too alone.

—Eben Moglen, *Anarchism Triumphant: Free Software and the Death of Copyright*

The dogma of monetary incentives, with which Moglen quarrels, is rooted in a philosophical history which reached its apogee in the work of Jeremy Bentham. According to his prescription, individuals act to attain that which is good for them—the useful. Bentham believed utility could be universally defined and function as a guide for social organization. Although this belief faded with the acceptance that utility was subjectively defined, henceforth the new utilitarians would elevate the market as decentralized arbiter and archive of the useful. With only minor modifications this model remains hegemonic today in the form of neoclassical economics. In its liturgy, *homo economicus* is a rational agent performing countless cost-benefit calculations, and undertaking instrumental action to reach logical ends. As Alain Caillé puts it:

"... one can characterize utilitarianism as all purely instrumental conceptions of existence, which organize life as a function of a calculation or a systematic logic of means and ends, in which an action is always carried out for some other purpose than itself, tied to the sole individual subject, supposedly closed in himself and sole master, addressee and beneficiary of his acts."

—Alain Caillé, *De L'anti-utilitarisme, 2006*

Much critical thought accepts substantial parts of the utilitarian perspective, finding in it a clear logic that allows both an understanding of the motor of human behavior, and the possibility to intervene in and transform it. Of course the explanatory power of this framework lies in its account of what people do to satisfy the ubiquitous need for money: food, clothing, shelter are just the least controversial of the needs usually offered only in return for payment. Failing congenital riches (or a social welfare state), one must earn money, and this need for income constitutes the main reason why most people work where they do: in contexts not freely selected by them.

Once the set of narrowly utilitarian explanations have been re-dimensioned as capturing just a subset of what drives action, their explanatory power can be more usefully harnessed.

Utilitarian belief was bolstered by studies of situations where the contingency of rewards on performance—incentives—does indeed increase/improve output. Social psychologists class such responsiveness to rewards—or threats—as evidence of 'extrinsic motivation'. But in recent years extensive research has demonstrated that monetary incentives can be counter-productive; over time they can 'crowd out' important non-monetary drives. Humans are not coin-operated machines, and need meaning, recognition and fulfillment.

Beyond the Carrot and the Stick

We do not do anything and everything solely for money. Even in the labor market we often trade off material gain against the reconciliation of other desires. These drives are classified by psychologists as 'intrinsic motivations'. The original definition comes from Deci:

> "One is said to be intrinsically motivated to perform an activity when one receives no apparent reward except the activity itself."
> —Deci 1971, p.105

Amongst the internal motivations identified as important are: the cultivation of self-determination (control over action); enhancement of self-esteem; performance of the self. What unites these elements is their importance in the formation of identity.

Deci's definition is problematic because it proposes a segregation of motivations which rarely exists; instead the quest for external reward and personal gratification co-exist in varying measures. This is true both for payed and voluntary labor (where it might be the quest for prestige). But intrinsic factors are plainly more important in voluntary pursuits, for the obvious reason that there is neither remuneration as compensation for doing something dissatisfying, nor monetary sanction for doing something badly.

Collaboration can provide a context where ability and commitment find acknowledgment, and a stage (especially online) on which we can choose to present ourselves in ways unavailable in the spaces of work or the local physical community. This concern with the self may help explain why attribution appears important to many participants. Because it is possible to get more done working with others, collaboration also offers the possibility of empowerment, and the chance to learn through interaction. On a more hedonistic level, these contexts also offer a space in which to play, unleash curiosity, engage in gestures of reciprocity and kindness, or indulge the sheer pleasure of sociality through participatory community.

Distinctions such as extrinsic and intrinsic, whilst schematic, help to parse and appreciate the different levels. The drives behind our actions are multifaceted, complex and resistant to exhaustive or exact dissection.

The Right Job...

"...you will probably be very much surprised when I say that there is really no such thing as laziness. What we call a lazy man is generally a square man in a round hole. That is, the right man in the wrong place."

—Alexander Berkman, *What is Anarchism?*

Implicit in all this is the fundamental importance of the nature of the task to be performed; motivations do not exist unanchored from specifics, but vary according to the nature and circumstances of what is to be done. Mundane, repetitive chores, or those with no graspable outcomes, are inherently less appealing. They neither challenge intelligence or catalyze learning, nor deliver a sense of making a meaningful contribution. On the other hand an objective that requires development of new skills, creativity, and that affords control over the nature and style of our contribution is more attractive. Surveys of the FLOSS community, for example, have repeatedly borne out that problem solving, pleasure in 'making things', and delight in the aesthetics of code are all attractive to contributors (Ghosh, Lakhani).

In his book, "Here Comes Everybody," Clay Shirky describes editing a Wikipedia entry on the fractal nature of snowflakes. He asks himself why he did it, and comes up with three answers. First he says it "was a chance to exercise some unused mental capacities—I studied fractals in a college physics course in the 1980s." The second reason is vanity: "the "Kilroy was here" pleasure of changing something in the world, just to see my imprint on it." The third motivation is simply "the desire to do a good thing. This motivation of all of them, is both the most surprising and the most obvious." (Shirky, p. 132)

... for the Right Person: Self-Selection

Where involvement is purely voluntary, people assume tasks which they feel suited to, or whose challenges attract them. This *self-selecting* character of participation distinguishes relations and activities in these projects from the mobilization of labour in the firm. As a result intrinsic motivations have a greater importance in these contexts, but this self-directed orientation is often alloyed with extrinsic motivations.

Online, the ease of sharing information erodes the obstacles between doing something for oneself and socializing it for others. As a software developer wrote in one weblog:

> "The fact is, in the Free Software world... the developer is the consumer. Applications are not programmed for some mythical "average consumer" but rather for real world applications. For example, if I need a new image viewer because none of the current ones (xv, ee, eog, etc.) meet my needs I simply sit down and program one to what I exactly want. Many companies already do this internally and create proprietary software. However, in the Free Software community that software which would otherwise be proprietary is made public.
>
> ... This is what I call productive selfishness. (Successful) Projects are created for their developers own personal reasons and are given to the community simply because someone else might have need of it and might want to extend it."
>
> —'Penguinhead', *Linux Today*

His formulation reminds us that the relationship between satisfying our own needs and helping others can often be an *encounter* rather than a collision.

Some free software programmers get paid to do what they love. For others money may not be immediately present, but reputation and prestige are; learning may be fun but it's also useful to guarantee ongoing employability; contact with others in a creative context may help appease self-doubt, but may also be the vector for the next job.

But in addition to these ambivalent convergences, there is always something more.

Interconnectedness

There is no individual universe available to inhabit: the responses, opinions and reactions of others are always in play and often important to us. Human interconnectedness is embedded, and manifests itself in the search for recognition, not the micro-celebrity of the online media, but rather a primary acknowledgment of being, dignity and worth.

Ultimately the motivations are myriad in a galaxy of online communities composed of a universe of subjectivities: forums filled with the argumentative, those looking for information and occasionally donating some themselves, the lonely seeking community, the vain, the depressed seeking support. Projects made of apprentices and mentors, practical people, proselytisers, entrepreneurial types, dreamers. In sum, a plurality, who one way or the other have built sustained communities.

7. Social Creativity

The reliance on the social in the field of economy is mirrored in the sphere of cultural production. Literature, music, film and art all draw from the pre-existing works which make up our mental universe. No creative act is strictly individual, relying as it does on the sources and styles of those which preceded it. Extensive legal and theoretical acrobatics have been required to keep this collectivity at bay, but in an era of digital production it is not clear that this will suffice.

Print Capitalism and Authorship: From Sovereign to Public

Literary historian Martha Woodmansee has chronicled how writing was understood as a derivative and collective enterprise during the middle ages and the renaissance. A book was seen as the product of many contributors, of which the writer was only one. Conceptualization of individual authorship, then, was a necessity imposed by the modalities of print capitalism and the need to establish clear property rights.

Prior to the first copyright law in 1710, the Statue of Anne, printing was regulated through a combination of self-policing by the Stationers Company (London publishers) and royal prerogative. The former took the form of a registry of who 'owned' specified titles, and members were expected to abide by its rules. The latter was expressed through awards of 'privileges', i.e. monopolies, over some books and fields by the crown. These structures were designed to keep peace within the trade and prevent the printing of seditious materials. Copyright law was created to achieve the same ends: to control and regulate dealings amongst publishers, and ensure accountability for that which was printed.

While the registry established rights of infinite duration, the law limited the term of exclusivity. Printers initially claimed a common law right in literary property, proposing that the term provided by copyright law was merely a statutory addition. In 1774 the courts rejected this claim of a perpetual common law right; this established the publishing industry as based on law rather than industry custom. The industry's pretensions to autonomy having been extinguished, and the sovereignty of the law established, copyright law began develop in correlation with the needs of the industry, encompassing new types of works (maps, music, etc) and granting longer terms of protection.

As industry and the reading public expanded, the justification for its monopoly grants changed. Under the influence of the eighteenth century romantic movement, copyright law began to be presented not as a means for intra-industry regulation but as a mechanism supporting the production of ideas by authors freed from the shackles of patronage. The free market would produce liberated writers, and their works would provide the public with enjoyment, knowledge and enlightenment.

To substantiate this new idea of authorship required that writing be recast as a unique record of the intellect behind it: property resided in the precise form of expression, above and beyond the physical artifact of the book itself and the ideas contained therein. Whilst property in this unique form belonged nominally to the author, it was transferred to the publisher. Publishers provided the means for authors to subsist, ergo, legal support for their needs was a means of supporting authorship. The rhetoric of the 'romantic author' provided a sympathetic figure whose creative genius could be mobilized to legitimate the need for a copyright monopoly. This formulation also glossed the inherent conflict between writers and publishers, and the asymmetrical power in their relationship; with rare exceptions, publishers held the whip.

In the US, British Empire, and throughout Europe, this idea that copyright represented a reward for genius or an incentive for the production of knowledge useful to the public became the official rationale for a monopoly grant.

Arcane as it may seem, this notion of authorship underwrites the logic of contemporary copyright law, and its assumptions are deeply implanted both in the functioning of the law and contemporary conceptions of creativity.

Copyright and the Collaboration Dilemma

The discourse was the way to create explicit property rights within the publishing system. But as commercial creativity became more complex with the development of the music, film and later software industries, such a framework risked encumbering business by generating too many rights-holders. Each of these works is obviously collective in manufacture, but allocation of rights to each participant would produce friction in terms of their trade on the market.

This risk has been dealt with by a slight of hand, whereby a corporate principal is ascribed as the creator and the producers are considered as agents of their will, otherwise known as the work for hire doctrine. In other cases this is dealt by a contractual transfer of any rights, as is the case for the recipients of patents in the corporate employment.

Copyright law recognizes only two other modalities of collaboration. A 'collective work' is one which contains several works each of which are copyrighted themselves; the US statue offers anthologies and encyclopedia as examples. Joint works on the other hand are 'prepared by two or more authors with the intention that their contributions be merged into inseparable or interdependent parts of a unitary whole'. In this case each author can license the result on a non-exclusive basis and must pay the other(s) their share. While this tallies with the nature of many online productive practices, it creates a massive rights thicket by spreading blocking rights extremely widely. The consequences of giving each participant rights comparable to those of individual authors is a situation of paralysis.

Collectivity Resurgent

This basic conception of creativity as individual leaves the legal framework ill-equipped to deal with contemporary forms of wide-scale cooperative production. Collectivity is inscribed in both their form and architecture, from the discursive and serial nature of problem solving in forums, to the version control histories of software and wikis. These practices are confronted with a legal framework unable to respond to their needs. This explains why so many have turned to alternative forms of copyright licensing which change copyright's defaults so as to facilitate or even encourage free collaboration, such as the GPL and (later) Creative Commons.

In addition to these artifacts native to the digital context, online activity generates copious amounts of documentary evidence of the collective nature of design and execution in every other field. As creative practices become more explicitly derivative and collaborative, the legal stability of copyright's categories are being strained past breaking point. Movements in all fields of the arts had foreshadowed these tensions. Practices of montage recycling of footage in cinema, collage, the cut-up in writing, re-photography all reflected the fact that in an age of ubiquitous media, creative reinterpretation would necessarily take the form of recombining, 'appropriating' pre-existing elements. Courts struggled incoherently with these challenges, ruling inconsistently and inventing progressively more peculiar distinctions. These practices were clearly not about 'piracy', but were in direct contradiction to the claims of original genius of the 'romantic author'. The result was chaos, but as long as access to the technology was restricted by high entry costs, it effected only a discrete group.

The spread of the personal computer and software for media manipulation in the 1990s, followed by the arrival of high speed domestic connectivity, washed away the final flood wall. Doctrines developed to regulate industrial cultural producers are in crisis, confronted by a public itself now equipped with the tools of production and distribution.

8. Open Relationships

Like romantic relationships, open collaborations are based on mutual trust, and trust alone can be too fragile a social fabric to support human interaction. Most romantic relationships base their trust in terms of sexual and emotional exclusivity, a contract that is socially accepted and helps both members of the relationship feel safe by agreeing to restrict their intimacy with others. It is a simple rule. Respecting that rule shows respect for the partner, both privately and socially; breaking that rule shows disrespect and can lead to social humiliation, pain, and nasty breakups.

Many find this convention dull, sexist, and restricting, but when eliminated - when the simplicity and the clarity of the contract is gone - the need to create new boundaries quickly follows. These transplanted borders establish new rules where that respect can manifest itself again. Those who refuse to do so find themselves single very quickly, or very frustrated.

In an open relationship a different social pact governs. Each couple decides their own rules, but they establish these rules so as to map out boundaries, and abide by them. These rules preserve the cohesiveness of the core relationship, prevent awkward or uncomfortable situations. Some agree to "never take your lover to our favorite restaurant" or "you two should never hang out with our mutual friends". Some rules regulate special times, such as "don't spend the night" or "don't celebrate birthdays" in order to keep the sense of exclusivity. Whether more rigid or more flexible, all of these rules serve the same purpose: to make sure nobody gets hurt and nobody feels cheated.

So, the traditional arrangement of sexual exclusivity simplifies the terms of romantic partnerships. In the non-romantic world, people avoid getting hurt or cheated in a collaboration by using a contract; in traditional work settings this contract is written down on paper, and signed, but in a less formal collaboration this is a social contract, an agreement or understanding.

Under a contract, the terms of collaboration are clear and legally binding. When collaboration is open and there is no explicit contract, the binding terms can be a shared passion, a common goal, a sense of community (or the lack thereof), but nevertheless, the need for implicit and explicit structure remains.

Depending on the specific collaboration any number of norms (either rigid rules or informal social practices) may need to be established to address the regulating issue. Generally healthy collaborative processes establish norms to cover behaviors relating to coordination, transparency, attribution, autonomy, generosity, respect and freedom of movement.

Glossary: **Contract = temporary contract**
(friendship and otherness)

A temporary contract is a virtual deal or document between people interacting or working with each other, actualized in a specific time (e.g. the duration of a project) to aid in initiating modes of communication and/or articulate possible positions taken within it. Perhaps we can here speak of a soft contract (without a legal document) that allows flexibility and shifting between these various entities/selves. Within a cultural assemblage (which is already contaminated by the law and governmental structures) is it advisable to avoid contractual language and vocabularies—or does this lead to tyranny (see *The Tyranny of Structurelessness*)? Can we speak of a social contract without reproducing vocabularies and strategies of legal/illegal bureaucracies? Can we transform them into valuable tools? Are there agreements that can advance a certain common set-up and strategy based on friendship, the affirmation of otherness, and selves constantly undergoing shifts and transformations?

What can friendship bring into questions of deals between selves, based on a "processual self creation" (Félix Guattari)? In his text 'Friendship as Community: From Ethics to Politics' Simon O'Sullivan (2004) notes: "[...] important is the involvement in what Guattari calls the 'individual-group-machine'—basically, interaction with others which allows for a process of *resingularisation*—in which individuals 'create new modalities of subjectivity in the same way an artist creates new forms from a palette' (Guattari, 1995)." For O'Sullivan "friendship as community [ultimately] has to be *lived*: one cannot produce Spinoza's common notions without experiencing joy—and one cannot, following Guattari, creatively produce one's subjectivity in isolation from others. If there is to be (as Hardt and Negri amongst others claim) a new society, it is not one that will arrive from 'out there', but one that will emerge from right here—from ourselves working on the stuff of our own lives." Following this logic, the temporary contract here becomes intrinsically linked with our lives, with the here and now, and with something that might escape regulation systems. At least it potentially precedes and blocks them.

9. Participation and Process

Our conception of what constitutes fair treatment vary according to context, as do our reactions to being treated unfairly. If someone jumps the queue in front of us in a shop, it's annoying but quickly forgotten. But if we contribute time or money to establishing a collective enterprise, and instead it is subverted for other ends, then we feel angry and betrayed.

Our expectations and emotional intensity vary according to the degree we feel ourselves invested in, and part of, a shared project. As the intensity increases so does a need for procedural guarantees, transparency, fairness in terms of the division of benefits and acknowledgment.

Where participation is limited to small or occasional contributions, we may not even want to be drawn into time-consuming discussions about goals and methods. Likewise where our involvement is driven purely by whims of gratification rather than any desire to attain distant objectives. NASA's ClickWorkers project asks users to identify and count craters on Mars, and the combined inputs allow them to rationalize use of their internal research resources. While it is impossible to guess all the motivations which drive people to contribute, it is obvious that no one expects to be able to actively influence NASA's overall agenda by contributing, nor to control the organization directly.

Sustained involvement requiring a substantial expenditure of effort, or active engagement to create or promote something deemed of worth or importance, demands a more careful framework. Care is required because participation implicates our sense of identity. Defection by others, a sense of betrayal, anger at manipulation or exploitation are destructive not only to the immediate project but to willingness to collaborate in the future. On the other hand every collaboration needs room also to change, and a breathing space which acknowledges the different levels of commitment of its participants, which themselves will vary over time.

While an explicit process is no panacea to the problems that arise when we deal and work with others, it can anticipate and mitigate the most damaging consequences when things go awry, whilst protecting the flexibility necessary to adapt.

Productive dissent

Any collaborative endeavor (paid, unpaid or underpaid as the case may be) risks becoming an echo chamber—a shared space in which participants operate from a similar world-view, mostly agree with each other, and quibble over details and technicalities without being about to raise larger, riskier questions about the work and why or whether it needs to exist.

Friendship, love and attachment can be crucial to collaborative work. With friendship comes responsibility and good friends can (should) challenge each other. But "open" collaborations are quite susceptible to the inadvertent suppression of dissent because of the convergence of like-minded thinkers and the lack of formal processes for recruitment of participants, voicing of disagreements, and raising of fundamental ontological issues (see the Interlude: 'Tyranny of Structurelessness' section).

Particularly if they are meant for an audience beyond the creators, collaborative projects should always take heed of their "outside" and engineer a process for productive antagonism. This might take the form of invitations to participate or partner, public forums, consultation over beer, training processes for newbies or provocative blog posts, among many other possible formats.

Glossary: **Dissent**

> Necessary as a constituting element for any debate, agency, educational turn, and (micro)community to emerge. In an effort to reduce the "echo chamber" effect, collective efforts need to engineer methods of collective listening as a means of gathering productive dissent.

Essentially, collaborative projects need to develop modes of collective listening to the environment outside their boundaries. Can collaborations learn to be more sensitive and receptive?

It is challenging to be sensitive and it is hard to listen. It may not come naturally. Hence it is an engineering problem—something to be purposefully designed.

Architect and theorist Markus Miessen eloquently interrogates the consensus-seeking rhetoric around calls for "participation", an insight that can be extended to collaboration. All too often a romantic view of "harmony and solidarity" is assumed. Miessen writes that he "would like to promote an understanding of conflictual participation, one that acts as an uninvited irritant". In the writing of this book we have seen first hand how easy it is to subsume dissent within a collaborative framework (because, in many ways, it takes more work to disagree, to communicate, to reconcile or part ways). Collaborations may actually encourage a sort of passivity; pre-existing projects have a logic and inertia that can be impossible for outsiders to crack, to truly intervene on, even if they are encouraged by originators to do so. In a way, collaboration produces subjects—one must submit to the project, become subjected to it—it order to engage.

Authority in Distributed Creation

"We reject: kings, presidents and voting.

We believe in: rough consensus and running code."

—David Clark, "A Cloudy Crystal Ball—Visions of the Future" (1992)

Online communities are not organized as democracies. The most accountable of them substitute a deliberative process of discussion for majority-based voting. This derives from the fact that the original initiative emanated from one or a couple of people, and because participants are there of their own volition. Majority rule is not seen as inherently good or useful. The unevenness of contributions highlights the fact that a disproportionate part of the work in a project is done by a smaller sub-group. Within a political sphere that privileges production this trends towards the valuing of ability and commitment, sometimes phrased in the language of 'meritocracy'. (That said, as we think about these structures and arrangements outside the software model it gets more tricky. How do we apply the concept of "meritocracy" to the realm of art making, where there are fewer objective standards? Yes, a person may be very proficient with video editing software, but that doesn't necessarily make them a "good editor." To continue with this example, there are no objective criteria for things like narrative sensibility, pace, style of cutting, and so on—only conventions and taste. What and who defines merit?)

Founders particularly have considerable power, derived from the prestige accruing from a successful project, recognized ability, and their network centrality—having had most opportunity to forge relations with newcomers, and an overview of the technical structures and history of the project. These factors give them authority. This hierarchical element is nonetheless diffused by the modular nature of productive organization: sensible structures devolve authority over their parts so as to maximize the benefits of voluntary contribution.

This architectural enabling of autonomy extends also to newer users, who can take initiative free from having to continuously seek permission and endorsement. However their contributions may not be incorporated if considered substandard or unnecessary, but such decisions arise out of a dialogue which must have some basis in efficiency, aesthetics or logic. Arbitrary dismissal of others in a community environment risks alienating others, which if generalized and persistent may place the whole edifice under strain, or even spark a fork or split.

Longstanding projects have also tended to give themselves defined legal forms at some point, thus the prevalence of foundations behind everything from Wikipedia to Apache. These structures often have charters, and sometimes hold elections to decide on the entry of the new members or appoint totemic figures.

Reputation and Trust

Influence derives from reputation—a substitute for trust in the online environment—which is accumulated and assessed through the use of persistent avatars, user names or real names. In addition to demonstrated aptitude, quantitative measures of commitment are also relied upon. Initial promotion of an editor's status on Wikipedia, for example, relies upon the length of time since the first edit, and the number of edits effected. Thereafter advancement also entails a qualitative evaluation of an editor's performance by their peers.

Higher user status allows the individual greater power over the technical tools that co-ordinate the system, and require confidence on the part of others that access will not abused. This threat is higher in software projects where hostile infiltration poses a real security risk given that the code will be publicly distributed. A variety of methods for vouching for each other are thus cultivated, new developers may require sponsors. In the case of Debian physical encounters between developers are used to sign each others' encryption keys, which are then used to authenticate the package management process, adding a further layer of robustness.

Process Fetishism

There's a risk of making a fetish of process over product, of the act of collaboration over the artifact that results from it. How important is it that a product was produced through an open, distributed network if, in the end, it serves the interests of the status quo? If it's just another widget, another distraction, an added value that some giant conglomerate can take advantage of, as in some cases of crowdsourcing? Does open collaboration serve a purpose or is it more like a drum circle, way more fun and interesting for the participants than for those who are forced to listen to it?

Collaboration is fundamental to human experience. It should be no surprise that collaboration also occurs online. The important question is what goals these new opportunities for cooperation and creation across space and time are put in service of.

In placing emphasis on process—on activity that is ongoing—there is also an implicit elevation of change over stasis, movement over stillness, speed over slowness, fluidity over fixity. 'Nowness', newness, and real time are heralded as superior to a long view. These attributes must be questioned as goods in and of themselves.

It is true that digital media needs to move, to be updated, to stay relevant —but we should pause to critically reflect on why this is necessarily the case. After all, hardware and software manufacturers, using the principles of "planned obsolescence" push consumers to buy new, supposedly improved devices every season. Or worse, they design them to malfunction, forcing consumers to purchase replacements, as Giles Sade's *Made To Break: Technology and Obsolescence in America* makes fascinatingly clear.

This mindset was humorously visible early one morning as two of us walked passed an Apple Store in New York City on the way to work on this text. A line of people holding iPhones wrapped around the block, all of them waiting in eager anticipation of—you guessed it—the revamped iPhone. We question whether this impatient acquisitiveness, this obsession with the "latest and greatest," is an attitude we want to export to other modes of creative production. In a society committed to growth at any cost—to more, newer, faster, bigger, better—it's tempting to assume that change is progress, when there is in fact no such guarantee. And so, while we believe that open collaboration can be a valuable paradigm in certain circumstances, it should not be mistaken for a panacea. Alongside collaboration we must carve out space for projects that are no longer upgraded or made "new and improved," that are no longer in process, and as such provide counterbalance to the tyranny of the new and the now.

 Glossary: Acceleration/Deceleration

> A mode of speed or slowness that describes a changing intensity of movement. Data as well as our lives seem to be measure-able by modes of movement. The still or fixed position is perhaps a way of indicating a counter-position in this constant claim for moving and taking part in the seamless flux. Becoming slow and still and silent is possibly a radical, risky, more complex and therefore more attractive act.

10. Limits of Participation

What follows is one of two short essays published to frame "Re:Group: Beyond Models of Consensus," an exhibit at Eyebeam Art and Technology Center in New York City that, through a series of installations, examined participation as a dominant paradigm structuring social interaction, art, activism, architecture, and the economy (the exhibit also served as the umbrella for the second incarnation of the *Collaborative Futures* book sprint). To honor their philosophical differences and avoid subsuming conflict in pursuit of consensus, the curators of Re:Group released diverging statements. We've included the second statement, which presents a forceful critique of "participationism" to highlight one strategy, uncompromising as it may be, for retaining the heterogeneity inherent in any collaboration.

These days everyone—individuals, corporations, governments and DIY punks—idealizes participation. Many believe that when horizontal structures of participation replace top-down mechanisms of control, hierarchy and authoritarianism, this will eliminate apathy and disenfranchisement. While we acknowledge that distributed systems are proven and powerful tools for dismantling certain monolithic structures, we question an unalloyed faith in participation. As co-curators of the show we fought the temptation to simply celebrate the subversive potential of networked collaborations. Instead, we sought to critically analyze the contours of this emergent ideology, and to re-evaluate refusal, non-engagement, antagonism, and disagreement as fundamental to a participatory framework.

We are all the time besieged to Participate! Choose! Vote! Share! Join! And Like! And yet, we are all, already, integrated into structures of participation (whether we "like" it or not). We worry that a veneer of engagement only obscures deep flaws in the participation paradigm. Too often, it seems, progressives believe that power operates exclusively from above, that command and control emanate from some centralized, closed authority. It is no wonder that many latch on to notions of openness, transparency, and participation as radical ends in themselves; however we must not fetishise process over product.

Participatory frameworks are not in and of themselves politically significant, nor is power limited to distant and impersonal structures. Power is diffuse and distributed, operating through us and on us; participation therefore can turn into a vector for dominant ideologies as easily as it can liberate.

If participatory frameworks are to have any meaningful political consequence or activist import, they must intervene on some object, to operate in service of an end. Conflict is a necessary result of such collaboration, and a key driving force within it. Current conversations around participation idealize harmony and unison, but we ask whether synthesizing perspectives and valorizing consensus might actually subsume dissenting viewpoints, through the tyranny of compromise and the rule of the lowest common denominator. From this view, we fear a disavowal of power rather than an honest discussion about it.

And so we pass on politesse, and draw a line in the sand. We aren't interested in raising questions, exploring models of participation or experiments in collaboration. We take a position: that participationism plagues us. More than dismantling or distributing power, we've invisibilized and extended it. An intervention is in order, and we offer practices and programming that contribute to this conversation: foregrounding the contours and boundaries inherent in participation, the contradictions and conflicts in a fruitful collaboration.

—Not An Alternative, 2010

What is collaboration anyway?

11. First Things First

Information technology informs and structures the language of networked collaboration. Terms like "sharing", "openness", "user generated content" and "participation" have become so ubiquitous that too often they tend to be conflated and misused. In attempt avoid this misuse with the term "collaboration" we will try to examine what constitutes collaboration in digital networks and how it maps to our previous understanding of the term.

Sharing is the First Step

User Generated Content and social media create the tendency for confusion between sharing and collaboration. Sharing of content alone does not directly lead to collaboration. A common paradigm in many web services couples identity and content. Examples of this include blogging, micro-blogging, video and photo sharing, which effectively say: "This is who I am. This is what I did." The content is the social object, and the author is directly attributed with it. This work is a singularity, even if it is shared with the world via these platforms, and even if it has a free culture license on it. This body of work stands alone, and alone, this work is not collaborative.

In contrast, the strongly collaborative Wikipedia de-emphasizes the tight content-author link. While the attribution of each contribution made by each author is logged on the history tab of each page, attribution is primarily used as a moderation and accountability tool. While most User Generated Content platforms offer a one to many relationship, where one user produces and uploads many different entries or media, wikis and centralized code versioning systems offer a many to many relationship, where many different users can be associated with many different entries or projects.

Adding a second layer

Social media platforms can become collaborative when they add an additional layer of coordination. On a micro-blogging platform like Twitter, this layer might take the form of an instruction to "use the #iranelections hashtag on your tweets" or on a photo sharing platform, it might be an invitation to "post your photos to the LOLcats group." These mechanisms aggregate the content into a new social object. The new social object includes the metadata of each of its constituent objects; the authors name is the most important of this metadata. This creates two layers of content. Each shared individual unit is included in a cluster of shared units. A single shared video is part of an aggregation of demonstration documentation. A single shared bookmark is included in an aggregation of the "inspiration" tag on delicious. A single blog post takes its place in a blogosphere discussion, etc.

This seems similar to a single "commit" to a FLOSS project or a single edit of a Wikipedia article, but these instances do not maintain the shared unit/collaborative cluster balance. For software in a code versioning system, or a page on Wikipedia the single unit looses its integrity outside the collaborative context and is indeed created to only function as a part of the larger collaborative social object.

12. Coordinating Mechanisms create Contexts

Contributions such as edits to a wiki page, or "commits" to a version control system, cannot exist outside of the context in which they are made. A relationship to this context requires a coordinating mechanism that is an integral part of the initial production process. These mechanisms of coordination and governance can be both technical and social.

Technical Coordination and Mediation

Wikipedia uses several technical coordination mechanisms, as well as strong social mechanisms. The technical mechanism separates each contribution, mark it chronologically and attribute it to a specific username or IP address. If two users are editing the same paragraph and are submitting contradicting changes, the MediaWiki software will alert these users about the conflict, and requires them to resolve it. Version control systems use similar technical coordination mechanisms, marking each contribution with a timestamp, a user name, and requiring the resolution of differences between contributions if there are discrepancies in the code due to different versions.

The technical coordination mechanisms of the Wiki software lowers the friction of collaboration tremendously but it doesn't take it away completely. It makes it much harder to create contributions that are not harmonious with the surrounding context. If a contribution is deemed inaccurate, or not an improvement, a user can simply revert to the previous edit. This new change is then preserved and denoted by the time and user who contributed it.

Social Contracts and Mediation

Academic research into the techno-social dynamics of Wikipedia shows clear emergent patterns of leadership. For example the initial content and structure outlined by the first edit of an article are often maintained through the many future edits years on. (A Kittur, RE Kraut; Harnessing the Wisdom of Crowds in Wikipedia: Quality through Coordination) The governance mechanism of the Wiki software does not value one edit over the other. Yet, what is offered by the initial author is not just the initiative for the collaboration, it is also a leading guideline that implicitly coordinates the contributions that follow.

Much like a state, Wikipedia then uses social contracts to mediate the relationship of contributions to the collection as a whole. All edits are supposed to advance the collaborative goal—to make the article more accurate and factual. All new articles are supposed to be on relevant topics. All new biographies need to meet specific guidelines of notability. These are socially agreed upon contracts, and their fabric is always permeable. The strength of that fabric is the strength of the community.

> "If you're going against what the majority of people perceive to be reality, you're the one who's crazy"
> Stephen Colbert
> <*www.colbertnation.com/the-colbert-report-videos/72347/july-31-2006/the-word---wikiality*>

An interesting example of leadership and of conflicting social pacts happened on the Wikipedia Elephant article. In the TV show *The Colbert Report* Stephen Colbert plays a satirical character of a right wing television host dedicated to defending Republican ideology by any means necessary. For example he constructs ridiculous arguments denying climate change. He is not concerned that this completely ignores reality, which he claims "has a Liberal bias".

On July 31st, 2006, Colbert ironically proposed the term "Wikiality" as a way to alter the perception of reality by editing a Wikipedia article. Colbert analyzed the interface in front of his audience and performed a live edit to the Elephants page, adding a claim that the Elephant population in Africa had tripled in the past 6 months.

Colbert proposed his viewers follow a different social pact. He suggested that if enough of them helped edit the article on Elephants to preserve his edit about the number of Elephants in Africa, then that would become the reality, or the Wikiality—the representation of reality through Wikipedia. He also claimed that this would be a tough "fact" for the Environmentalists to compete with, retorting "Explain that, Al Gore!"

It was great TV, but created problems for Wikipedia. So many people responded to Colbert's rallying cry that Wikipedia locked the article on Elephants to protect it from further vandalism.
<*http://en.wikipedia.org/wiki/Wikipedia:Wikipedia_Signpost/2006-08-07/Wikiality*> Furthermore, Wikipedia banned the user Stephencolbert for using an unverified celebrity name (a violation of Wikipedia's terms of use <*en.wikipedia.org/wiki/User:Stephencolbert*>.

Colbert and his viewers' edits were perceived as mere vandalism that was disrespectful of the social contract that the rest of Wikipedia adhered to, thus subverting the underlying fabric of the community. Yet they were following the social contract provided by their leader and his initial edit. It was their own collaborative social pact, enabled and coordinated by their own group. Ultimately, Wikipedia had to push one of its more obscure rules to its edges to prevail against Stephen Colbert and his viewers. The surge of vandals was blocked but Colbert gave them a run for the money, and everyone else a laugh, all the while, making a point about how we define the boundaries of contribution.

13. Does Aggregation Constitute Collaboration?

Can all contributions coordinated in a defined context be understood as collaboration? In early 2009 Israeli musician Kutiman (Ophir Kutiel) collected video clips of hobbyist musicians and singers performing to their webcams posted on YouTube. He then used one of the many illegal tools available online to extract the raw video files from YouTube. He sampled these clips to create new music videos.He writes of his inspiration,

> "...Before I had the idea about ThruYou I took some drummers from YouTube and I played on top of them—just for fun, you know. And then one day, just before I plugged my guitar to play on top of the drummer from YouTube, I thought to myself, you know—maybe I can find a bass and guitar and other players on YouTube to play with this drummer."
>
> Kutiman on the ThruYou project
> <www.radiowroclove.pl/?p=151>

The result was a set of 7 music-video mashups which he titled "ThruYou —Kutiman Mixes YouTube". Each of these audiovisual mixes is so well crafted it is hard to remind yourself that when David Taub from NextLevelGuitar.com was recording his funk riff he was never planning to be playing it to the Bernard "Pretty" Purdie drum beat or to the user *miquelsi*'s playing with the theremin at the Universeum, in Göteborg. It is also hard to remind yourself that this brilliantly orchestrated musical piece is not the result of a collaboration.

When Kutiman calls the work "ThruYou" does he mean "You" as in "us" his audience? "You" as in the the sampled musicians? Or "You" as in YouTube? By subtitling it "Kutiman mixes YouTube" is he referring to the YouTube service owned by Google, or the YouTube users who's videos he sampled?

The site opens with an introduction/disclaimer paragraph:

> "What you are about to see is a mix of unrelated YouTubevideos/clips edited together to create ThruYou. In Other words—what you see is what you get.
>
> Check out the **credits** for each video—you might find yourself.
>
> PLAY >"
>
> <www.thru-you.com> (emphasis in the original)

In the site Kutiman included an "About" video in which he explains the process and a "Credits" section where the different instruments are credited with their YouTube IDs (like tU8gmozj8xY & 6FX_84iWPLU) and linked to the original YouTube pages.

The user *miquelsi* did share the video of him playing the Theremin on YouTube, but did not intend to collaborate with other musicians. We don't even know if he really thought he was making music: it is very clear from the video that he doesn't really know how to play the Theremin, so when he titled his video "Playing The Theremin" he could have meant playing as music making or playing as amusement. It would be easy to focus on the obvious issues of copyright infringement, and licensing, but the aspect of Kutiman's work we're actually interested in is the question of intention.

Is intention essential to collaboration?

It seems clear that though these works were aggregated to make a new entity, they were originally shared as discrete objects with no intention of a having a relationship to a greater context. But what about works that are shared with an awareness of a greater context that help improve that context, but are not explicitly shared for that purpose?

Web creators are increasingly aware of "best practices" for search engine optimization (SEO). By optimizing web pages, creators are sharing objects with a strong awareness of the context in which they are being shared, and in the process they are making the Google Pagerank mechanism better and more precise. Their intention is not to make Pagerank more precise, but by being aware of the context, they achieve that result. Although reductive, this does fit a more limited definition of collaboration.

The example of Pagerank highlights the questions of coordination and intention. Whether or not they are optimizing their content and thus improving Pagerank, web content publishers are not motivated by the same shared goal that motivates Google and its share holders. These individuals do coordinate their actions with Google's out of their own self interest to achieve better search results, but they don't coordinate their actions in order to improve the mechanism itself. The same can be said about most Twitter users, most Flickr users, and the various musicians that have unintentionally contributed to YouTube's success and to Kutiman's ThruYou project.

Collaboration requires Goals

There are multiple types of intentionality that highlight the importance of intent in collaboration. The intentional practice is different from the intentional goal. Optimizing a web page is done to intentionally increase search results, but unintentionally contributes to making Google Pagerank better. When we claim that intention is necessary for collaboration, we really are talking about intentional goals. Optimizing your site for Google search is a collaboration with Google only if you define it as your personal goal. Without these shared goals, intentional practice is a much weaker case of collaboration.

14. Collaborationism

As collaborative action can have more than one intent, it can also have more than one repercussion. These multiple layers are often a source of conflict and confusion. A single collaborative action can imply different and even contrasting group associations. In different group context, one intent might incriminate or legitimize the other. This group identity crisis can undermine the legitimacy of collaborative efforts altogether.

Collaboration with the enemy

In a presentation at the *Dictionary of War* conference at Novi Sad, Serbia in January 2008, Israeli curator Galit Eilat described the joint Israeli/Palestinian project *Liminal Spaces*:

> "...When the word "collaboration" appeared, there was a lot of antagonism to the word. It has become very problematic, especially in the Israeli/Palestinian context. I think from the Second World War the word "collaboration" had a special connotation. From Vichy government, the puppet government, and later on the rest of the collaborations with Nazi Germany.
>
> Galit Eilat, Dictionary of War video presentation "
> <dictionaryofwar.org/concepts/Collaboration_(2)>

While there was no doubt that *Liminal Spaces* was indeed a collaboration between Israelis and Palestinians, the term itself was not only contested, it was outright dangerous.

I remember one night in 1994 when I was a young soldier serving in an Israeli army base near the Palestinian city of Hebron, around 3:30am a car pulled off just outside the gates of our base. The door opened and a dead body was dropped from the back seat on the road. The car then turned around and rushed back towards the city. The soldiers that examined the body found it belonged to a Palestinian man. Attached to his back was a sign with the word "Collaborator".

Context and conflict

This grim story clearly illustrates how culturally dependent and context-based a collaboration can be. While semantically we will attempt to dissect what constitutes the context of a collaboration, we must acknowledge the inherit conflict between individual identity and group identity. An individual might be a part of several collaborative or non-collaborative networks. Since a certain action like SEO optimization can be read in different contexts, it is often a challenge to distill individual identity from the way it intersects with group identities.

> "The nonhuman quality of networks is precisely what makes them so difficult to grasp. They are, we suggest, a medium of contemporary power, and yet no single subject or group absolutely controls a network. Human subjects constitute and construct networks, but always in a highly distributed and unequal fashion. Human subjects thrive on network interaction (kin groups, clans, the social), yet the moments when the network logic takes over—in the mob or the swarm, in contagion or infection—are the moments that are the most disorienting, the most threatening to the integrity of the human ego."
>
> The Exploit: A Theory of Networks
> *by Alexander R. Galloway and Eugene Thacker*

The term "group identity" itself is confusing as it obfuscates the complexity of different individual identities networked together within the group. This inherent difficulty presented by the nonhuman quality of networks means that the confusion of identities and intents will persist. Relationships between individuals in groups are rich and varied. We cannot assume a completely shared identity and equal characteristics for every group member just by grouping them together.

We cannot expect technology (playing the rational adult) to solve this tension either, as binary computing often leads to an even further reduction (in the representation) of social life. As Ippolita, Geert Lovink, and Ned Rossiter point out

> "We are addicted to ghettos, and in so doing refuse the antagonism of 'the political'. Where is the enemy? Not on Facebook, where you can only have 'friends'. What Web 2.0 lacks is the technique of antagonistic linkage."
>
> The Digital Given—10 Web 2.0 Theses
> by Ippolita, Geert Lovink & Ned Rossiter
> <*networkcultures.org/wpmu/geert/2009/06/15/the-digital-given-10-web-20-theses-by-ippolita-geert-lovink-ned-rossiter/*>

The basic connection in Facebook is referred to as friendship since there is no way for software to elegantly map the true dynamic nuances of social life. While friendship feels more comfortable, its overuse is costing us richness of our social life. We would like to avoid these binaries by offering variation and degrees of participation.

15. Criteria for Collaboration

Collaboration is employed so widely to describe the methodology of production behind information goods, that it occludes as much as it reveals. In addition, governments, business and cultural entrepreneurs apparently can't get enough of it, so a certain skepticism is not unwarranted. But even if overuse as a buzzword has thrown a shadow over the term, what follows is an attempt to try and construct an idea of what substantive meaning it could have, and distinguish it from related or neighboring ideas such as cooperation, interdependence or co-production. This task seems necessary not least because if the etymology of the word is literally 'working together', there is a delicate and significant line between 'working with' and 'being put to work by'...

Some products characterized as collaborative are generated simply through people's common use of tools, presence or performance of routine tasks. Others require active coordination and deliberate allocation of resources. Whilst the results may be comparable from a quantitative or efficiency perspective, a heterogeneity of social relations and design lie behind the outputs.

The intensity of these relationships can be described as sitting somewhere on a continuum from strong ties with shared intentionality to incidental production by strangers, captured through shared interfaces or agents, sometimes unconscious byproducts of other online activity.

Consequently we can set out both strong and weak definitions of collaboration, whilst remaining aware that many cases will be situated somewhere in between. While the former points toward the centrality of negotiation over objectives and methodology, the latter illustrate the harvesting capacity of technological frameworks where information is both the input and output of production.

Criteria for assessing the strength of a collaboration include:

Questions of Intention

Must the participant actively intend to contribute, is willful agency needed? Or is a minimal act of tagging a resource with keywords, or mere execution of a command in an enabled technological environment (emergence), sufficient?

Questions of Goals

Is participation motivated by the pursuit of goals shared with other participants or individual interests?

Questions of (self) Governance

Are the structures and rules of engagement accessible? Can they be contested and renegotiated? Are participants interested in engaging on this level (control of the mechanism)?

Questions of Coordination Mechanisms

Is human attention required to coordinate the integration of contributions? Or can this be accomplished automatically?

Questions of Property

How is control or ownership organized over the outputs (if relevant)? Who is included and excluded in the division of the benefits?

Questions of Knowledge Transfer

Does the collaboration result in knowledge transfer between participants? Is it similar to a community of practice, described by Etienne Wenger as:

> "...groups of people who share a concern or a passion for something they do and learn how to do it better as they interact regularly."

Questions of Identity

To what degree are individual identities of the participants affected by the collaboration towards a more unified group identity?

Questions of Scale

Questions of scale are key to group management and have a substantial effect on collaboration. The different variables of scale are often dynamic and can change through the process of the collaboration. By that changing the nature and the dynamics of the collaboration altogether.

- Size—How big or small is the number of participants?
- Length (time)—How long or short is the time frame of the collaboration?
- Speed—How time consuming is each contribution? How fast is the

decision making process?

- Space—Does the collaboration take place over a limited or extended geographic scale?
- Scope - How minimal or complex is the most basic contribution? How extensive & ambitious is the shared goal?

Questions of Network Topology

How are individuals connected to each other? Are contributions individually connected to each other or are they all coordinated through a unifying bottle-neck mechanism? Is the participation network model highly centralized, largely distributed, or assumes different shades of decentralization?

Questions of Accessibility

Can anyone join the collaboration? Is there a vetting process? Are participants accepted by invitation only?

Questions of Equality

Are all contributions largely equal in scope? Does a small group of participants generate a far larger portion of the work? Are the levels of control over the project equal or varied between the different participants?

16. Continuum Set

The series of criteria outlined above provide a general guide for the qualitative assessment of the cooperative relationship. In what follows, these criteria are used to sketch out a continuum of collaboration. The following clusters of cases illustrate a movement from weakest to strongest connections. This division is crude, as it sidelines the fact that within even apparently weak contexts of interaction there may be a core of people whose commitment is of a higher order (e.g. ReCaptcha).

The Weakest Link...

(1) Numerous technological frameworks gather information during use and feed the results back into the apparatus. The most evident example is Google, whose PageRank algorithm uses a survey of links between sites to classify their relevance to a user's query.

Likewise ReCaptcha uses a commonplace authentication in a two-part implementation, firstly to exclude automated spam, and then to digitize words from books that were not recognizable by optical character recognition. Contributions are extracted from participants unconscious of the recycling of their activity into the finessing of the value-chain. Web site operators who integrate ReCaptcha, however, know precisely what they're doing, and choose to transform a necessary defense mechanism for their site into a productive channel of contributions to what they regard as a useful task.

(2) Aggregation services such as delicious and photographic archives like flickr, ordered by tags and geographic information, leverage users' self-interests in categorizing their own materials to enhance usability. In these cases the effects of user actions are transparent. Self-interest converges with the usefulness of the aggregated result. There is no active negotiation with the designers or operators of the system, but acquiescence to the basic framework.

(3) Distributed computing projects such as SETI and Folding@Home require a one-off choice by users as to how to allocate resources, after which they remain passive. Each contribution is small and the cost to the user is correspondingly low. Different projects candidate themselves for selection, and users have neither a role in defining the choice available nor any ongoing responsibility for the maintenance of the system. Nonetheless the aggregated effect generates utility.

Stronger...

(4) P2P platforms like BitTorrent, eDonkey and Limewire constitute a system where strangers assist one another in accessing music, video, applications, and other files. The subjective preferences of individual users give each an interest in the maintenance of such informal institutions as a whole. Bandwidth contributions to the network guarantees its survival, and promises the satisfaction of at least some needs, some of the time. Intention is required, especially in the context of attempts at its suppression through legal action and industry stigmatization. Links between individual users are weak, but uncooperative tendencies are mitigated by protocols that require reciprocity or bias performance in favour of generous participants (eg BitTorrent, emule).

(5) Slashdot, the technology related news and discussion site is extraordinary in not actually producing articles at all. Instead stories are submitted by users and then filtered. Those published are either selected by paid staff, or voted on by the user-base. Following this, the stories are presented on the web page and the real business of Slashdot begins: voluminous commentary ranging from additional information on the topic covered (of varying levels of accuracy), to analysis (of various degrees of quality), to speculation (of various degrees of pertinence), taking in jokes and assorted trolling along the way. This miasma is then ordered by the users themselves, a changing subset of whom have powers to evaluate comments, which they assess for relevance and accuracy on a sliding scale. The number and quality of comments presented is then determined by users themselves by configuring their viewing preferences. User moderation is in turn moderated for fairness by other users, in a process known as metamoderation.

In addition to the news component of the site, Slashdot also provides all users with space for a journal (which predates the blog), and tools to codify relations with other users as 'friends' or 'foes' (predating and exceeding Facebook). The system behind the site, Slashcode, is free software and is used by numerous other web communities of a smaller scale.

(6) Vimeo, a portal for user-produced video, shelters a wide variety of sub-cultures/communities under one roof. Two characteristics distinguish it from other apparently similar sites: the presence of explicit collective experimentation and a high level of knowledge sharing. Members frequently propose themes and solicit contributions following a defined script, and then assemble the results as a collection.

Several channels are explicitly devoted to teaching others techniques in film production and editing, but the spirit of exchange is diffuse throughout the site. Viewers commonly query the filmmaker as to how particular effects were achieved, equipment employed, etc. The extent to which Vimeo is used for knowledge sharing distinguishes it from Youtube, where commentary regularly collapses into flame wars, and brings Vimeo close to Wenger's concept of a "community of practice" (see the above quote from Etienne Wenger in 'Questions of knowledge transfer').

Vimeo is nonetheless a private company whose full time employees have the final word in terms of moderation decisions. Nonetheless the community flourishes on a shared set of norms which encourage supportive and constructive commentary, and a willingness to share know-how in addition to moving images.

...Intense

(7) Although there is something of an over-reliance on the Wikipedia as an example, its unusually evolved structure makes it another salient case. The overall goal is clear: construction of an encyclopedia capable of superseding one of the classical reference books of history.

The highly modular format affords endless scope for self-selected involvement on subjects of a user's choice. Ease of amendment combined with preservation of previous versions (the key qualities of wikis in general) enable both highly granular levels of participation and an effective self-defense mechanism against destructive users who defect from the goal.

At the core of the project lies a group who actively self-identify themselves as *wikipedians*, and dedicate time to developing and promoting community norms especially around the arbitration of conflicts. Jimmy Wales, the project's founder, remains the titular head of wikipedia, and although there have been some conflicts between him and the community, he has in general conceded authority, but the tension remains without conclusive resolution.

(8) FLOSSmanuals, the organization that facilitated the writing of this text you are reading, was originally established to produce documentation for free software projects, a historically weak point of the FS community. The method usually involves the assembly of a core group of collaborators who meet face to face for a number of days, and produce a book during their time together.

Composition takes place on an online collective writing platform called booki, integrating wiki like version history and a chat channel. In addition to those physically present, remote participation is solicited. When focused on technical documentation, the functionality of the software guides the shape of the text. Where conceptual, as in the case of the current work, it is necessary to come to an agreed basic understanding through discussion, which can jumpstart the process. Once underway both content and structure are continually edited, discussed and revised. On conclusion the book is made freely available on the web site under a CC license, and physical copies are available for purchase on-demand.

(9) Closed p2p communities for music, film and text, such as the now suppressed Oink, build archives and complex databases. These commonly contain technical details about file quality (resolution, bit-rate), illustrative samples (screenshots), relevant additional information (imdb links, track listing, artwork), descriptions of the plot/director/musician/formal significance of the work.

In addition most sites have a means of coordinating users to ensure persistence of data availability. If someone is looking for a file currently unseeded, preceding downloaders are notified, alerting them to the chance to assist. When combined with the fixed rules of protocol operation and community specific rules, such as ratio requirements (whereby one must upload a specified amount in relation to the quantity downloaded), there is an effective mechanism to encourage or oblige cooperation. Numerous other tasks are assumed voluntarily, from the creation of subtitles, in the case of film, to the assembly of thematic collections. All users participate in carrying the data load, and a significant number actively source new materials to share with other members, and to satisfy requests.

(10) Debian is constructed around a clearly defined goal: the development and distribution of a gnu/linux operating system consistent with the Debian Free Software Guidelines. These guidelines are part of a wider written 'social contract', a code embodying the project's ethics, procedural rules and framework for interaction. These rules are the subject of constant debate. Additions to the code base likewise often give rise to extended discussion touching on legal, political and ethical questions. The social contract can be changed by a general resolution of the developers.

Debian exemplifies a 'recursive community' (see Christopher Kelty, 'Two bits'), in that they develop and maintain the tools which support their ongoing communication and labour. Developers have specified tasks and responsibilities, and the community requires a high level of commitment and attention. Several positions are appointed by election.

17. Non-Human Collaboration

It is interesting to ask ourselves if humans are the only entities which might have agency in the world. Do you need language and consciousness to participate? In her lecture "Birth of the Kennel," Donna Haraway has observed that "It isn't humans that produced machines in some unilateral action—the arrow does not move all in one way [...] There are very important nodes of energy in non-human agency, non-human actions." <*www.egs.edu/faculty/donna-haraway/articles/birth-of-the-kennel*>

Even further, Bruno Latour suggests it might be possible to extend social agency, rights and obligations to automatic door closers, sleeping policemen, bacteria, public transport systems, sheep dogs and fences. Taking this view perhaps we might begin to imagine ourselves as operating in collaboration with a sidewalk, an egg-and-cheese sandwich, our stomachs, or the Age of Enlightenment.

Most of our conversations about collaboration begin with the presumption of a kind of binary opposition between the individual and social agency. Latour solves this problem by suggesting that there are actor-networks —entities with both structure and agency. We ignore the non-human at our own peril, for all manner of non-human things incite, provoke, participate in and author actions in the world.

How might it inform and transform our conversations about collaboration if we imagined ourselves to be collaborating not only with people but with things, forces, networks, intellectual history and bacteria?

Case Studies

18. Boundaries of Collaboration

Collaboration can be so strong that it generates hard boundaries. Boundaries can intentionally or unintentionally exclude the possibility to extend the collaboration. Potentially conflict can also occur at these borders.

For example, Book Sprints often develop strong and lasting collaborative relationships centered around the production and maintenance of a book. The intense social environment of a sprint can produce sharp borders around the collaboration. While Book Sprints produce texts that are available on an open license, and within a technical mechanism that allows for remote contributions, this does not in itself collapse the border between the sprint group and those 'outside' of the room.

Glossary: *Speed*

Speed or Velocity. The supposed opposite to slowness. Interesting in relation to the (book) sprint, the experience of time, and the possible value we can attribute to interrupting the constant stream of data and information. Be slow. Different from progress which claims a "future" or some kind of "utopian dimension", speed remains more abstract in the sense of not necessarily indicating a particular direction. Remember Paul Virilio on questions of speed in his book "Speed and Politics: An Essay on Dromology", 1977 [1986]; Dromology—'Dromos' from the Greek word to race (Virilio 1977:47). Meaning: the 'science (or logic) of speed'. Read Hiroshi Yoshioka's 'The Slowness of Light' <*www.iamas.ac.jp/~yoshioka/SiCS/e-text/en_published_040331_slownessoflight.html*> in relation to high-speed technology, where he suggests not opposing speed, but using it for different purposes.

In a recent Book Sprint for the "Google Summer of Code Mentoring Guide", a group of very experienced Free Software developers (each were also experienced GSoC mentors) collaboratively wrote a book in two days. The collaboration was fluid and intense and generated a very useful text which has since been propagated throughout the GSoC community. Some weeks later a freelance technical editor with free time offered to copy edit the book but the group rejected the offer to collaborate. The reasons for this exclusion were complex, but discussion centered around the group feeling uncomfortable for reasons ranging from 'not knowing' the person, to issues about attribution, ownership and quality control.

Excluding potential collaborators in this scenario was intentional and considered by the group to be entirely appropriate. The group felt this was fully consistent with the ideals of Free/Open Content in that freely licensed content does not require compulsory collaboration, it has the potential to enable it, and the group felt that if others wanted to work on the text they were free to fork the text and create their own version.

While it is possible to discuss the groups decision about who they collaborate with, there are also consequences to this exclusivity that must be considered. In this case study it is interesting to note that since the rejection of the offer no work has been done on the 'shared resource', and hence the product has not been maintained. In other words, as a result of hard exclusionary boundaries all collaborative activity eventually ceased.

19. P2P : The Unaccepted Face of Cultural Industry?

Napster's launch and popularization in 1999 transformed the unauthorized online circulation of media from an activity of adepts to a mass phenomena. In its aftermath, the public discourse ramped up, pivoting around the denomination and vilification of P2P as 'piracy'. But this language was nothing new; it represented a quantitative expansion and escalation in a longer struggle over cultural production. Campaigns against home taping and the litigation against the VCR are just two instances in that history.

The precarious legal situation of P2P communities make them difficult to discuss. Legal campaigns have propelled a bifurcation into two species: public and private. The former, exemplified by the Pirate Bay with a user base of millions, are predominantly (all though not solely) used for the distribution of mainstream content. 'Oink' exemplifies a private community: dedicated to sharing music, it was closed by police in autumn 2007 and had 180,000 members at the time of its death. The site's founder was subsequently tried and acquitted on charges of conspiracy to defraud; four site administrators were charged and convicted of copyright infringement.

What is of interest here however is the collaborative productivity of these communities, and the way in which their growth has produced a new method of distribution for works more generally.

A New Type of Archive

Private P2P communities are generating new types of archives based on federating the sum of their users' contributions. Whereas media collections in the analogue period were private libraries, in the digital they coagulate into decentralized archives, whose communities assemble databases of metadata and information about the items shared. Sub-communities form around genres or individual producers, seeking out lost works, and aggregating literature and other material considered important to contextualize their presentation.

In the case of film communities, large numbers of works are translated and subtitled by participants *ab initio*. Networked cinephilia is also about attempting to extend, or create, an audience for undervalued films by reintroducing them via virtual channels. The low cost of digital distribution enables the release of works preciously regarded as too economically marginal to justify commercial release. Here we are generally not speaking about works belonging to Hollywood or major record labels. P2P community participants capture recordings from television and radio or digitize existing analogue versions for distribution. In many cases, there is no possibility to effectively access these works through normal channels, and this functions as both an attraction and a justification, as users feel that their activity is not parasitic, but rather assists in the preservation of otherwise abandoned works.

This activity is in part driven by the proselytizing instincts of enthusiasts, but is also nurtured by the technical framework within which such sites function. Users are expected both to take things they like and to reciprocate by making their own contributions. This expectation is enforced through the application of ratio rules, which limit the amount one can download in proportion with the amount uploaded. Ratio functions both as a proxy for currency and as a reputation index. Due to their contested legality, access to these communities is normally by invitation only, and invitations are provided only to those who have demonstrated a willingness to observe the rules of the game by maintaining an acceptable ratio.

Hackers affiliated with these communities develop software required to accomplish many of the tasks outlined above: subtitle authoring and synchronization; transcoding; tools for working with audio and video streams; database management; improved distribution protocols and clients.

Lastly the availability of large amounts of unrestricted digital works enable further downstream creation, by making available materials which can be recycled into new works, be they fan-films, criticism, pastiche or parody. Barriers to participation in this subsequent layer of production are diminished through the provision of images and music in readily manipulable form.

Glossary: **Non-Documents**

The obsession and notoriety to document a process, a thinking and action, produce an eruption of belief in materialities. It is important to emphasize that outside the Western context there is not necessarily such a desire to build a mega-archive. Can we learn to non-document, or to create more virtual documents in the form of brain images, memory data, and partially return to oral histories? How would that affect our relation to the elderly and socially (technologically) excluded people?

Independent Channels for Distribution

It may have been the lure of free or rare goods that initially motivated users to install p2p clients, but in the process something new has been created: an unowned distribution infrastructure. Until the emergence of p2p networks, serving large amounts of media content was expensive, posing a dilemma for independent producers who wanted to distribute works online. Media enterprises invest large amounts of resources, and employ expensive service companies, to ensure rapid data delivery to viewers and finesse the user experience, but the costs involved are beyond the reach of most independents. Video platforms such as Youtube and Vimeo provide such facilities for serving video, but exact control in terms of framing, content 'suitability', or commercial exploitation - advertising - in return.

The installed base of p2p clients, dispersed amongst the population, allows independents to shift the burden of data transfer to fans and supporters. Large file-sharing sites such as the Pirate Bay and Mininova periodically donate promotional space to encourage users to download works made available voluntarily by their producers. In so doing, a sort of diffuse 'channel' is created, which has no-one point of entry, as anyone can use their web-site to disseminate links to items available in these networks, be it on Bit Torrent, eMule, Gnutella etc. Once injected into these networks the work is effectively impossible to censor, although its level of visibility, and the performance of the network as a distribution platform, will vary according to the level of user mobilization and enthusiasm behind it.

20. Anonymous Collaboration II

Tor, The Onion Router, is a Free Software tool that makes internet use anonymous by effectively hiding your IP address. After installing the software, your computer becomes a node on the TOR network, sending encrypted packets of data from node to node until they arrive at their final destination. The data travels through so many nodes in the network that it obscures the path to the original IP address. If you send an email, your data will be encrypted from your computer to the last computer prior to its destination. This final computer on the network is the IP address that will be reported in any network analysis, and this IP addresses is any IP in the network other than yours.

Tor was originally designed for the U.S. Navy to protecting government communications. It resists traffic analysis, eavesdropping, and any nosy activity, from both in and outside the onion network. It is a very convenient software, widely available and easy enough to use.

Technically, Tor hides you among the other users on the network. While the level of practical commitment is low—it just requires a connection and downloading some code—the personal investment is high: by using Tor, you are part of a community of computer users that help each other hide from state and corporate control mechanisms. Strangers help you defend your privacy, avoid censorship and grant you a degree of personal freedom by fooling surveillance mechanisms with a mirrors trick. Hiding in this way is illegal in some countries. And, what is more interesting, you don't know who these strangers are.

The reason why Tor is so important is not because of what it does, it is because of what it represents. In the Tor forest, everybody covers for everybody, but nobody knows who the others are. They are not friends, and they are not family. It is *that* anonymous. Anyone can use Tor to do things other people wouldn't approve, like downloading porn or attacking other people's computers. Or things that governments would not approve of, like posting dissident information, the most notorious case being a very active chap called Wikileaks.

It doesn't have to be heroic; maybe you just want to browse the most milquetoast sites on the Internet with complete privacy. By using Tor, you join a bunch of strangers in declaring everybody has the right to complete privacy and collaborate anonymously to grant yourself and others that constitutional right.

21. Problematizing Attribution

> "I get credit for a lot of things I didn't do. I just did a little piece on packet switching and I get blamed for the whole goddamned Internet, you know? Technology reaches a certain ripeness and the pieces are available and the need is there and the economics look good—it's going to get invented by somebody."
>
> Paul Baran

A few years ago, the unofficial fanclub website of a very popular Spanish band became notorious for reasons beyond their commitment to the band. As is customary, the site included a page with all the lyrics from all the songs recorded by the band over the years, listed in chronological and alphabetical order. But, at the end of the page, the girls in charge claimed copyright to the whole content under their own names! After a while, a disclaimer note appeared after the copyright terms. The disclaimer explained what they thought it was obvious: as they were the first website dedicated to the band, they have had to copy all the lyrics from the CD booklet to the website by hand.

Clearly, it was a big task.

From their perspective, they result of transcribing the material rightly belonged to them, the same way the CD belonged to the major company who sold it. In those terms it was clearly no fair for other fans to just go and copy-paste the lyrics on their own websites. On the other hand, they pointed out that "we can't stop anyone from copying the lyrics from the record, just as we did".

Never before have creators and companies been so concerned about intellectual property. Courts are full of musicians accusing other musicians of stealing parts of their compositions. J.K. Rowling has sued and has been sued for stealing the plot and characters of her wildly successful Harry Potter books. Two years ago, Adidas won the exclusive right to use parallel stripes in groups of two, three and four. They stopped at four, but only because K-swiss has been using the five stripe logo since 1966.

All these infamous cases make no sense at all. Is it strange then that four teenagers grow confused about what belongs to who when it comes to their favorite band? Independent from any economical ambition, we have a mostly hilarious confusion of a claim of ownership (copyright) with a demand for acknowledgment, to be credited for the useful task they felt they had performed (attribution). Of course their act was in itself a copyright infringement, because someone 'owned' exclusivity in the lyrics after all. But then again, they "stole" from their favorite band in order to promote them even further. And the confusion never stops.

If only all cases were so naive. The inflation of copyright claims has been so radical, and the mismatch with the contemporary usage seems so dramatic, that people are inspired to get in on the game even when the law gives them no grounds. Additionally this example highlights how attribution is often conflated with ownership.

22. Asymmetrical Attribution

The New York Times Special Edition project was a great success. This collaboratively made knock-off of the New York Times was dated July 4, 2009, months into the future from the morning of November 12, 2008 when 100s of volunteers distributed these papers on the streets of New York City. The 14 page perfect replica contained all the news that the creators of the newspaper hoped to print, including the end of the war in Iraq, the arrival of universal health care, and a new maximum wage law. The paper was fake, but at first glance it caught its readers in a moment of belief. Though the news was obviously impossible, it was convincing because it was so well crafted, and so realistically handed out on the streets by volunteers wearing New York Times aprons. Bogs went crazy, the project sped through the national and international press and copies of the paper immediately appeared on eBay as collectors items.

The project was conceived by two to four people, organized by a group of 10, created by an even larger group of 50, and distributed by hundreds of others on the streets of New York. It was by all accounts a successful collaboration. An internal conflict over leadership amongst the group of 10 organizers resulted in one person leaving the group; this is not unusual, and not the focus here. The organizers worked tirelessly for months leading up to the day of the event, managing the team of people creating the newspaper and the companion website.

The first outlet to cover the event was Gawker. The first Gawker post that appeared cited the location of the main distribution van. Once Gawker writer Hamilton Nolan realized this was breaking news, they did some more research, and found one of the organizational emails describing the planned event. These emails were being circulated amongst a private, but fairly open group. The emails were not signed, and they were from an as-yet unknown domain, becausewewantit.org, that was purchased simply as a cover to distribute those emails from; it was allowed to lapse, and is now squatted by an advertiser. Gawker matched the IP address in the long header of the email to other IP addresses of the activist duo The Yes Men, and updated the post attributing the authorship to The Yes Men. Hamilton Nolan wrote

> "The email address that sent out this message was linked to the site of The Yes Men, longtime liberal prank group that has been doing things just as complex and finely tuned as this for years. The Yes Men run the Because We Want It site, through which they set up this prank. They wanted to be anonymous for a while allegedly, but too late."
> <gawker.com/5084164/fake-new-york-times-declares-iraq-war-over-heres-who-did-it>

And from then on, the project authorship was assigned to The Yes Men. The group of organizers sent out a press release later in the day from the email address "New York Times Special Edition <special[at]nytimes-se.com>". Nowhere in the email is attribution given, or authorship claimed. Rather, inquiries are directed to "writers@nytimes-se.com." But as that press release spread across the Internet it was referred to as a Yes Men press release. Even the New York Times itself fell into this pattern in one of their several articles on the New York Times Special Edition, stating that "On Wednesday, the Yes Men issued a statement about the prank," and linking to this appearance of the press release: *<www.poynter.org/forum/view_post.asp?id=13699>* *<cityroom.blogs.nytimes.com/2008/11/12/pranksters-spoof-the-times/>*.

The server was in fact a Yes Men server, and one of the Yes Men was one of the project originators and key organizers, but the most important factor here is that the collective had no way to define their own identity in the face of the powerful media coverage that had pinned it to a known entity. One of the Yes Men was a central organizer, but it wasn't "a Yes Men project". It was a project by a large coalition of pretty well known artists and activist groups.

And then the news started emailing special@nytimes-se.com for interviews. From a group of 50, how do you choose a representative? This is always a problem when a project gets major attention. Who gets interviewed? Who represents the project at festivals? Who receives the awards, if there are awards? In most instances with this project, Andy Bichlbaum and Steve Lambert were the representatives. They were the two of the four who had originally conceived the project that carried it through to completion, raising the funds, and coordinating the massive team of volunteers. In their CNN interview they repeatedly emphasized that the project was conceived, organized, and executed by a large group of people, and that they are there as representatives of that larger group *<www.youtube.com/watch?v=dO6Oi3XUYgg>*. Steve Lambert's website documents the project and lists every single volunteer and group that worked on or sponsored the project *<www.visitsteve.com/work/the-ny-times-special-edition/>*. And yet, Gawker's rushed attribution still sticks to the project, highlighting the problems in contesting representation amidst massive asymmetries of broadcast power.

Here the problem arises when a collaboratively produced project are 'privatized' through their representation by individuals. How can such impositions be prevented or, at least, limited? In this case, at the outset every effort was made to not make this a project of The Yes Men, but society and the media at large is so preoccupied with assigning authorship that the first question Gawker wanted to know was "Who made this?" As the event was unfolding they found an answer that was satisfactory enough for them, and that incorrect answer became the story that was told from that point onwards.

This scenario raises a number of questions. One problem this highlights is that ownership of URLs and servers often equates into ownership of projects: So who registers the URL, and who maintains the server? But the larger question is how do you negotiate attribution in a collaboration where there are significant imbalances in power: Different collaborators have different media presences. And how do you negotiate attribution when there are many organizers, and many collaborators, who are working on something that is almost certain to achieve a large degree of impact?

23. Can Design By Committee Work?

In the excited celebration of the merits of Free Software and radically networked production methods, an important truth is left unspoken. Networked collaboration shines at the low levels of network protocols, server software and memory allocation, but has consistently been a point of failure when it comes to user interfaces. How come the methods that transformed code production and encyclopedia writing fails to extend into graphic and interface design?

What follows is an investigation into the difficulties of extending the FLOSS collaboration model from coding to its next logical step, namely interface design. While it dives deep into the practical difference between these two professional fields, it might also serve as a cautionary tale to consider before prematurely declaring the rise of "Open Source Architecture", "Open Source University", "Open Source Democracy"…

The Challenges

Scratching an Itch

Coders are fulfilling their own need to change software, to make it their own. They might have diverging motivations but if you're already modifying something for yourself it is really easy to answer the "why share?" question with "why not?". By the time the code executes correctly the immediate users of the software, the coders themselves, are already familiar with the software and can operate it even without a delicately crafted user interface.

Therefore the motivation to take an extra step and to invest in a usable interface that would extend the user base beyond the original geek-pool is not obvious. This is already working for me, so what itch am I scratching when I work hard to make it usable by others who can't help me code it?

And for designers themselves, what is the incentive to make the design process more collaborative? Will others make my design better? Would they be able to better communicate my thoughts than I can?

Beyond that, FLOSS interface design suffers from a chicken and egg problem; Most designers don't use FLOSS tools, and so they are not aware that they could make the software better. As a result FLOSS often suffers from inferior interfaces that makes designers shy away from it and stick to their proprietary tools. The cycle continues…

Granularity

Both software and wikis are made of granular building blocks, namely the character. This makes every typo an invitation to collaborate. My first Wikipedia edit was a typo correction, the second was an additional reference link, the third was actually a whole paragraph and that lead to more substantial contributions like adding a whole new article and so on.

Each granular step gets you closer to the next granular step. This ladder of participation makes every next step easier. It also allows easy comparison of changes which provide transparency, accountability, moderation and an open license to try and possibly fail knowing that you can always revert to the previous version.

You don't have that with design as the changes are not granular and are not as easily traceable. The first step is steep and a ladder is nowhere to be found.

Encoding / Decoding

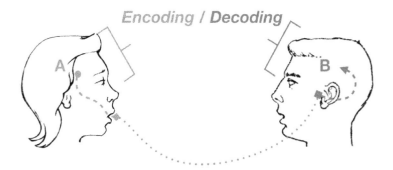

In his 1980 article "Encoding / Decoding" cultural theorist Stuart Hall defines communication in terms of code. To describe it briefly let's imagine a spoken conversation between Alice and Bob. Alice encodes her framework of knowledge into the communicable medium of speech. Assuming Bob can hear the sounds and understand the spoken language, he then decodes the sounds into a framework of knowledge.

Both encoding and decoding are creative processes. Ideas are transformed into messages that are then transformed into ideas again. The code Alice uses for encoding is different than the one used by Bob for decoding. Alice could never just telepathically "upload" ideas into Bob's brain. We would all agree that is a good thing.

Let's entertain Hall's ideas of encoding and decoding in software. Alice is an

FLOSS hacker, Bob is collaborating with her as a designer. Alice is writing software code, she knows when it executes and when it doesn't as the program communicates that through error messages. When she is happy with the result she uploads the code to an online repository.

Bob then downloads the code to his computer and since it executed on Alice's computer, it also executes on his. When Alice and Bob collaborate through programming language they are literally using the same code for encoding and decoding.

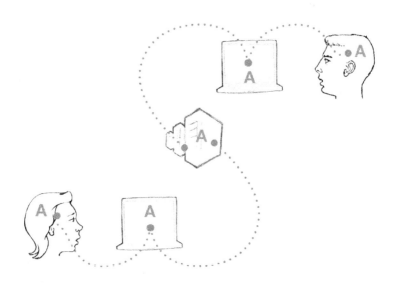

Alice always chooses one of her three favorite programming languages. Being a designer, to communicate a message visually Bob starts by defining a visual language—graphics, color, layout, animation, interaction... If Alice or any other developer had to reinvent a new programming language on every single project we would not be speaking about FLOSS now.

Bob needs to define a graphic language, a standard for the collaboration. Doing that is already a major part, possibly the most important part of the creative work. Whoever works with Bob will need to accept and follow these standards, relinquish control and conform to Bob's predefined graphic language. These artificial constraints are harder to learn and conform to than the constraints of a programming language. While constraints and standards in technology are the mother of creativity, in design they can often feel artificial and oppressive.

Beyond that, within a collaboration, when Bob tries to argue for the merits of his design, unlike in the case of Alice's code he cannot prove that it executes flawlessly, or that it is faster or more resource efficient. The metrics are not as clear.

It is important to remember, in collaboration on code Alice and Bob have a third collaborator, one that cannot be reasoned with - the computer. This collaborator will simply not execute anything that doesn't fit its way of work. On the other hand, as long as it is syntactically correct and satisfies the inflexible collaborator even "ugly code" executes and muddles through. And so, the different voices expressed in code are flattened into a single coherent executed application.

For better or worse, we lack this inflexible collaborator in design. It doesn't care about our communicative message and it doesn't level the playing field for communicative collaboration. And so, the different voices in design simply spell inconsistent multiplicity that dilutes the communicative message.

One might turn to Wikipedia as a testament to successful non-code-based collaboration, but Wikipedia enforces very strict and rational guidelines. There is no room for poetry or subjectivity within its pages.

So is it simply impossible?

Not necessarily. If we step out of the technical construct of the FLOSS methodology we can identify quite a few networked collaborations that are transforming and often improving the design process.

It is tempting to see free culture and the free sharing of media as evidence of collaboration, but the availability of work for remixing and appropriation does not necessarily imply a collaboration. Sharing is essential to collaboration but it is not enough.

WordPress, the leading Free Software blogging tool is an interesting example. Looking to redesign the WordPress administration interface, Automattic, the company leading the Wordpress community has hired HappyCog, a prominent web-design firm. And indeed in March 2008, WordPress 2.5 launched with a much improved interface. Through a traditional design process HappyCog developed a strong direction for the admin interface. Eight months later Automattic released another major revision to the design that relied on HappyCog's initial foundations but extended them far beyond.

One of the interesting methodologies used to involve the WordPress community in the design process was a call for icon designers to provide a new icon set for the interface. Within two weeks six leading icon sets were up for vote by the community. But rather than just asking for a blanket like/dislike vote, they were invited to provide detailed assessments of consistency, metaphor coherence and so on. Some of the icons designers ended up acknowledging the superiority of other contributions and voting against their own sets. But the final icon set was indeed a collaborative effort, as some of the icons were altered based on inspiration from the other sets.

Another example is the evolution of grid systems for web design. Half a century after the rise of Swiss style graphic design, some design bloggers suggested some of its principles could apply to web design. Those suggestions evolved into best practices and from there into Blueprint CSS, an actual style sheet framework. The popularity of that framework inspired other frameworks like 960.gs and others. Similar processes happen in interaction design as well, with the pop-up window evolving into elegant lightbox modules and then being repeatedly changed and modified as open source code libraries.

Other design minded experiments in Free Software like the ShiftSpace platform challenge the web interface power structure. ShiftSpace allows users to interact with websites on their own terms by renegotiating the interface and proposing different interactions on top of the page. Projects like ShiftSpace aim to expand the limited participatory paradigm of the web beyond user generated content to also include user generated interfaces.

Make it happen!

There are ways to make Open Source design work without falling into the traps often characterized as "design by committee". We are already seeing **designers scratching their own itches** and contributing creative work to the commons.

Lecturing designers (or any users) and demanding they use bad tools for ideological reasons is counter productive. Designers often use free tools (or make unauthorized use of proprietary tools) only because they are free as in free beer. So to win any new user, Free Software should be pitched on the full range of its merits rather than ethics alone. While the ethics of "free as in free speech" are very convincing for those who can "speak" code, for those who do not have the skills to modify it the openness of the source is not such a compelling virtue.

Free Software tools have won on their broad merits many times, not only on the low-level system and network fronts. Wikis and blogging software are interaction and communication tools that have been invented by the Free Software community and have maintained a lead over their proprietary competitors. Networking and collaboration are the bread and butter of Free Software, and these are advantages the community should leverage.

In the same way that Wikipedia extends the Free Software collaboration model by leveraging the **granularity of the character**, so can design. From a collaboration standpoint, where possible, it is preferable to use code to implement design (like HTML, CSS). Beyond that, collaborators should adopt distributed version control systems for both code and image files. Rather than trying to compete with proprietary software by creating open clones, the Free Software community can leverage its experience as an advantage and focus on new collaborative paradigms for version control and collaboration.

Encoding / Decoding

Glossary: *Bike-Shedding*

"Why Should I Care What Color the Bikeshed Is?

The really, really short answer is that you should not. The somewhat longer answer is that just because you are capable of building a bikeshed does not mean you should stop others from building one just because you do not like the color they plan to paint it. This is a metaphor indicating that you need not argue about every little feature just because you know enough to do so. Some people have commented that the amount of noise generated by a change is inversely proportional to the complexity of the change."

<*www.freebsd.org/doc/en_US.ISO8859-1/books/faq/misc.html#BIKESHED-PAINTING*>

Finally, There are ways for us to better analyze the **encoding and decoding** of the communicative message. We can formalize processes of **collaborative encoding**. We can start by conducting networked design research using existing research tools, that way we might come up with design decisions collaboratively. We can define modular and extensible languages that embody the design decisions and still allow for flexibility and special cases (like Cascading Style Sheets). We should also learn how to document these design decisions we take so they serve the rest of the collaborators. Designers have been doing it for many years in more traditional and hierarchical design contexts, compiling documents like a brand book or a design guide.

For the **decoding** part, we should realize that many design patterns are rational or standardized and can leverage a common-ground without compromising the creative output. For example underlined text in a sentence on the web almost always implies a hyperlink. We can choose to communicate a link otherwise but if we try to use this underline styling as a sort of emphasis (for example) we can expect users will try to click on it.

User experience research, technical aspects of design, best practices in typography, icon use, interaction paradigm, these are all aspects of design that can be researched and assessed according to measurable parameters. Thorough research of these can provide a basic consensus for shared expectations of how a message will be interpreted. A lot of this work is already taking place on design blogs that over the past few years have been publishing a lot of research on the subject.

Finally, the substantial parts of design that still cannot be easily quantified or assessed on a unified rational ground, should be managed through trust and leadership. In the absence of any convenient meter of coding meritocracy, a resilient community of practice must develop design leadership whose work and guidance is respected and appreciated.

Scaling Subjectivity

It comes down to the deep paradox at the heart of design (interface, architecture, product...). We are trying to create a subjective experience that would scale up—a single personal scenario that can be multiplied again and again to fit a wide array of changing needs by a vast majority of users. The thing is subjectivity cannot be scaled, that's what makes it subjective, and therefore the attempts to create a one size fits all solution are bound to fail, and the attempts to customize the solution to every individual user in every individual use case are also bound to fail.

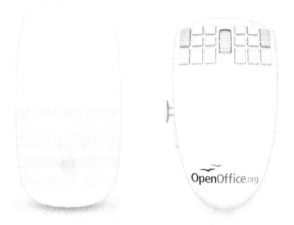

Chris Messina gives a great example of this paradox in a comparison between Apple's Magic Mouse and the Open Office Mouse <factoryjoe.com/blog/2009/11/07/open-source-design-and-the-openofficemouse/>. While Apple's solution is a slick and clean one-button device, the OOMouse has "*18 programmable mouse buttons with double-click functionality; Analog Xbox 360-style joystick with optional 4, 8, and 16-key command modes; 63 on-*

mouse application profiles with hardware, software, and autoswitching capability;" and more... While Apple's Magic Mouse embodies the company's commitment to design leadership at the price of user choice, the OOMouse embodies the Free Software's preference of openness and customization over unified leadership.

Successful FLOSS projects have always benefited from a mix of the two approaches, a combination of openness and leadership. Finding a similar nuanced approach in other fields is required if we ever hope to extend the FLOSS model beyond code. We cannot just sprinkle the pixie dust on everything and expect wonders. This applies also to design. But hopefully we can make some progress by demystifying the process, make sure we apply collaboration wisely when it does makes sense, and come up with new ways when it doesn't.

24. Multiplicity and Social Coding

The Linux kernel, arguably one of the most important FLOSS projects, was managed without a version control system until 2002. Linus Torvalds, the project leader, disliked centralized version control systems, which he considered unsuitable for kernel development. The Linux kernel is a very large and complex software project, has extraordinary quality demands and also attracts thousands of developers. Changes were meticulously tracked through a distributed hierarchy of delegates, but the system was showing strain.

In 2002 Linus finally decided that a "distributed version control system" (DVCS) would match the project's needs. The Linux kernel was migrated to the proprietary BitKeeper version control system, a selection which sparked great controversy because of its closed license. In 2005, licensing disputes eventually led to the creation of freely licensed distributed version control system and the DVCS named Git was created.

Distributed Version Control Systems operate on a different model than repositories managed by a centralized, client-server system. The DVCS model is peer-to-peer and, while it can be configured to resemble traditional client-server transactions, it can also support more complex interactions. In a DVCS system, every developer works locally with a complete revision history, and changes can be pushed and pulled from any other peer repository. The version control system has vastly improved support for merging across multiple repositories, and all working checkouts are effectively forks, until they are merged back onto a canonical trunk.

The demands on the Linux kernel project prefigured demands on other projects. In the past few years distributed version control systems have dramatically increased in popularity. Mercurial, Bazaar, and Git have emerged as the most popular open source DCVS systems, and hosting services have launched offering each of these systems free of charge for open source projects. Google Code began supporting Mercurial, alongside Subversion, repositories in mid 2009 (Paul, 2009b). Cannonical, the company which sponsors the Ubuntu GNU/Linux distribution, offers free Bazaar hosting to open source projects on Launchpad.net. In February 2008 GitHub.com launched, a "social coding" site which provides Git hosting and rich social networking tools to all developers using the site, gratis for FLOSS code. Bitbucket.org offers similar social networking tools around Mercurial, and describes itself as "leading a new paradigm of working with version control".

The centralized hosts of peer-to-peer protocols broker a new balance between centralization and federation. They facilitate coordination, but do not mandate it. A site like GitHub can track and aggregate multiple branches of development, but branching does not require any permission or upfront coordination. Instead of requiring an upfront investment of attention and energy to coordinate development activities, DVCS concentrate on improving the mechanisms for developers to track, visualize, and merge changes. The costs of coordinating collaboration is deferred, and the communication overhead required to synchronize and align different branches of code is (hopefully) reduced.

There is a fascinating culture emerging around DVCS, facilitated by software, but responding to (and suggesting) shifts in collaboration styles. As one developer explains:

> "SourceForge is about projects. GitHub is about people... A world of programmers forking, hacking and experimenting. There is merging, but only if people agree to do so, by other channels... GitHub gives me my own place to play. It lets me share my code the way I share photos on Flickr, the same way I share bookmarks on del.icio.us. Here's something I found useful, for what it's worth... Moreover, I'm sharing my code, for what it's worth to me to share my code... I am sharing my code. I am not launching an open source project. I am not beginning a search for like minded developers to avoid duplication of efforts. I am not showing up at someone else's door hat in hand, asking for commit access. I am not looking to do battle with Brook's Law at the outset of my brainstorm"
> —Gutierrez, 2008

Sometimes developers simply want to publish and share their work, not start a social movement. Sometimes they want to contribute to a project without going through masonic hazing rituals. DVCS facilitates these interactions far more easily than traditional centralized version control systems and the hierarchical organizations which tend to accompany them. Part of what makes this all work smoothly are very good tools to help merge disparate branches of work. This all sounds chaotic and unmanageable, but so did concurrent version control when it first became popular.

 Glossary: **Distribution**

An inquiry into the channels and formats that disperse, publicize and create audiences for ideas, objects, and data. Online / offline / at the thresholds. Distribution via a particular channel is not always the end point. Distributed material may possibly be adapted, modified, hacked, remixed and re-distributed by a user or group. This modification may or may not be legal depending on intellectual property and licensing issues. But the user might do it anyway.

In anecdotal accounts of switching to DVCS, developers describe an increase in the joy of sharing—the tools help reduce the focus on perfecting software for an imagined speculative use and the overhead of coordinating networks of trusted contributors. The practice really emphasizes the efficient laziness of agile programming, and helps people concentrate on the immediate requirements, rather than becoming preoccupied with endless planning and prognostications.

In some respects, this emerging style of collaboration is more freewheeling than an anonymously editable wiki, since all versions of the code can simultaneously exist—almost in a state of superposition. Wikis technically support the preservation of diversity in a page's history, but in a centralized wiki the current page is the (ephemeral) final word. DVCS are developing richer interfaces to simultaneously represent diversity, and facilitate the cherry picking of features from across a range of contributors. The expression of a multiplicity of heterarchical voices is explicitly encouraged, although there is a hidden accumulation of technical debt that accrues the longer a merger of different branches of work is delayed. Of course, sometimes you may actually want to start a community or social movement around your software, and that is still possible but is now decoupled, and needs to be managed with purposeful intent.

The Present

25. Crowdfunding

Tanda is a Mexican example of a system of collectively funding large purchases or creating rotating credit associations. It originated in Puebla, Mexico around 1899, and is said to have been inspired by a similar system brought by Chinese immigrants. These systems exist in many cultures. The core principle behind the Tanda is that every week or month everyone in the Tanda contributes a set amount of money. Each time, that money is all given to one person in the Tanda. Each time it rotates. The Tanda is used to make large purchases that would otherwise require formal loans, use as see money to start businesses, or to pay for important infrastructural improvements for communities *<www.anthro.uci.edu/html/Programs/Anthro_Money/Tandas.htm>*.

An online success

When Apple announced that their new computers were switching to the Intel chipset, aficionados immediately speculated about whether these new computers would be able to run Windows. On January 22, 2006 OnMac.net put up a page to collectively raise a bounty for the first person to write a bootloader to solve the problem. If the problem wasn't solved, the money was to be donated to the EFF. Over $20,000 was raised, on March 13 someone using the handle narf2006 posted the solution *<forum.onmac.net/showthread.php?t=64>*.

The effort was successful in that it not only achieved its goal of creating intense competition to accomplish an extremely challenging task that many people wanted accomplished, it also forced Apple three weeks later to release their still buggy, but much more stable version.

It is worth noting that almost all of the principles of collaboration had been followed: there was a clear goal and organization, everyone was properly attributed, the process was transparent, and the trust was maintained throughout. Even the names of the donors to the project are still upon the homepage, just as the project promised *<www.everymac.com/articles/q&a/windows_on_mac/faq/xom-hack-for-running-windows-on-mac.html>*.

An online failure

Fundable.com was launched in 2005, and promised to create a crowdfunding platform. Apparently many people were able to use it successfully, but many others had significant problems. These problems exploded when prominent author Mary Robinette Kowal had a terrible experience where Fundable held the donated money, neither disbursing it, nor refunding it to its original donors. This blew up on a BoingBoing.net post and, shortly thereafter, the company shut down.

The story behind the breakdown is the story of a failed collaboration. One side of the failure was detailed by one of the two partners on the Fundable.com homepage when he took it off-line October 1, 2009 <*files.spontaneousderivation.com/fundable-capture/index.html*>.

The details of who said what when, and who did what when, are fairly irrelevant, as the one-sided account is, well... one-sided. What is fundamentally clear is that the project failed because the collaboration between the two partners failed. There was a total breakdown in communication, trust, transparency, etc <*www.maryrobinettekowal.com/journal/my-very-bad-experience-with-fundable-com/*> <*boingboing.net/2009/08/22/fundable-rips-off-hu.html*>.

Kickstarter

Kickstarter.com has taken up this concept of crowdfunding with what seems to be significant initial success. The premise is simple: an individual defines a project that needs funding, defines rewards for different levels of contribution, and sets a funding goal. If that pledges meet the funding goal, the money is collected from pledgers, distributed to the project creator, who uses the funding to make the project. If the project does not reach the funding goal by the deadline, no money is transferred. Most projects aim for between $2,000 and $10,000.

Kickstarter pledges are not donations, as most of the contributions are associated with tangible rewards, nor are they a form of micro-venture capital, as funders retain no equity in the funded project. While crowdfunding need not limited in topic, Kickstarter is focused almost exclusively on funding creative and community focused projects. Part of their goal is to create a lively community of makers who support each other. At the end of their first year, they gave out a number of awards including the project with the most contributors, the project that raised the most money, and the project that reached their goal the fastest, but the award that might be most telling is for the "Most Prolific Backer":

"Jonas Landin, Kickstarter's Most Prolific Backer, has pledged to an amazing 56 projects. What motivates him? "It feels really nice to be able to partially fund some one who has an idea they want to realize."

<blog.kickstarter.com/post/318287579/the-kickstarter-awards-by-the-numbers>

One curious conundrum arose when Diaspora sought only to raise $10,000 to develop an open source social networking platform ended their campaign with $200,642. <www.kickstarter.com/projects/196017994/diaspora-the-personally-controlled-do-it-all-distr> Their fundraiser came at the same time as a wave of Facebook privacy roll-backs, perfectly matching the simmering discontent with Facebook to their privacy focused project. This enormous success has created a high level of public scrutiny that has led to public complaints about a number of aspects of the project, including the openness of their development process. <identi.ca/conversation/32668503> Though this is not the place to discuss the relative merits of these process-based critiques, it is worth noting that this might be an example of too much of a good thing. The four collaborators only asked for $10,000 to work on this project in lieu of a summer internship, but ended up with twenty times that amount. They also ended up bearing the weight of the hopes, desires, and scrutiny that came with that funding. The most successful FLOSS projects tend to be developed in obscurity; few, if any prior FLOSS project been developed in this kind of fishbowl. It is still to be seen if the success of Diaspora's crowdfunding has set them up for expectations they cannot live up to, or if it has set the stage for the adoption of the platform they are creating.

Funding the New York Times Special Edition

As previously mentioned, a post dated knock-off of the New York Times was distributed by hundreds of volunteers. They printed a newspaper based on a wish list of news. As the motto states, they printed "All the news we hope to print", a twist on the NY Time's famous phrase "All the news that's fit to print".

In order to fund the printing and distribution of the newspaper, the anonymous organizers came up with a campaign that emphasized the hope embodied in the newspapers' mission and retained the projects anonymity. They sent out an open call to thousands of people to donate to a large secret project to build a better world, without a clear description of what was being proposed.

As vague as it sounds, people sent more than ten thousand dollars in small donations simply based on a simple idea of optimism and hope. This model was effective in motivating a base for change, and was tapping into desire for change that co-existed in the Obama election campaign.

But is Crowdfunding Collaboration?

Crowdfunding clearly works as a means for democratized patronage. Kickstarter and others may demonstrate that providing a platform for crowdfunding is a good business. One question is how far crowdfunding can scale in terms of size of projects supported and platform businesses required.

It seems unclear whether crowdfunding itself *is collaboration*. However, it seems that crowdfunding can abet collaboration in at least two respects. First, its efficient democratic patronage may result in more projects retaining the freedom to collaborate rather than being restrained by the boundaries of an institution or business that may not be permeable to outside collaborators. Second, crowdfunding could lead to increased engagement between putative creators and consumers in a way that changes, mixes, and makes collaborative both roles.

26. Ownership, Control, Conflict

> *"Free cooperation has three definitions: It is based on the acknowledgment that given rules and given distributions of control and possession are a changeable fact and do not deserve any higher objectifiable right. In a free cooperation, all members of the cooperation are free to quit, to give limits or conditions for its cooperative activity in order to influence the rules according to their interests; they can do this at a price that is similar and bearable for all members; and the members really practice it, individually and collectively."*
> —Christoph Spehr, *The Art of Free Cooperation*

Ownership

In the industrial information economy the outputs are owned by the company that produces them. Copyrights and patentable inventions produced by employees are transferred to the corporate owner, either directly or via the 'work for hire' doctrine, which treats the acts of individuals as extensions of the employers will.

Collaborations amongst small groups often use similar proprietary methods to control the benefits arising from their outputs, but parcel out ownership amongst collaborators. In larger-scale settings, however, the very concept of ownership is turned on its head: where proprietary strategies seek to exclude use by others, these approaches prevent exclusion of others from using.

In the traditional workplace, the labor relationship was set out in a legally binding manner, whose terms were clear although imbalanced. In the digital era however the distinction between the time and space of work and that of play is ambivalent, and those dedicating their energy are often not employees. Licenses play a role partially analogous to that of the labor contract. Cases where volunteer contributions were subsequently privatized, such as Gracenote's enclosure of the Compact Disk Database (CDDB), have demonstrated the risks inherent in not confronting the ownership question (the takeover and commercialization of the IMDB is another less dramatic example).

The two principal licensing schemes used in free software and free culture production today reflect this. The GNU Public License, stewarded by the Free Software foundation, guarantees the rights to use, distribute, study and modify, *provided* the user agrees to abide by the same terms with any downstream outputs. Creative Commons (CC) licenses are more diverse, but that most commonly employed within large scale collaborations, the BY-SA (Attribution and ShareAlike) license, functions in the same manner. However, amongst individual users and small-team production there continues to be wide use of CC licenses, which permit distribution but bar commercial use.

Conflict

This licensing approach creates a system where rich repositories of data artifacts are available for reuse, also commercially, for those who abide by the rules: commons on the inside, property to the outside. Following Spehr, we can see that this strategy of preventing exclusive ownership allows anyone disagreeing with the direction taken by a project to leave *without* having to sacrifice the work that they have invested, because they can bring it with them and take up where they left off.

Such configurations are useful for a second reason. In traditional proprietary organizations the disgruntled have three options: *exit*, *loyalty*—putting up with it-, or *voice*—speaking out in opposition (the terminology is Albert Hirschman's). Because speaking out often incurs awkward conflicts and the possibility of stigmatization or expulsion, it is heavily disincentiv. Once the power of ownership is contained, however, one can leave the collaboration without abandoning the project, and the pressure to withhold criticism and disagreement is correspondingly attenuated. This can encourage conflicts to be played out in a potentially useful manner within a project, and makes exit an act of last resort.

Although this licensing protects participants' access to the outputs, there is always a cost to leaving: loss of any recognition and visibility already attained, technical infrastructure, and the consumption of energy through acrimony.

Forking and Merging

There have been many successful software forks over the years, demonstrating that the guarantees actually work. In some cases the second project supersedes the original, in others a period of separation is sufficient to cool tempers and reconcile differences, culminating in a reunification around new terms.

Glossary: *Fork*

1. As a piece of cutlery or kitchenware, a fork is a tool consisting of a handle with several narrow tines (usually two, three or four) on one end. The fork, as an eating utensil, has been a feature primarily of the West.
   <*en.wikipedia.org/wiki/Fork*>
2. (software) When a piece of software or other work is split into two branches or variations of development. In the past, forking has implied a division of ideology and a split of the project. With the advent of distributed version control, forking and merging becomes a less precipitous, divisive action.
   <*en.wikipedia.org/wiki/Fork_%28software_development%29*>

The disruptive force of forking is greater in an environment whose default is to maintain code in centralized, collaboratively maintained repositories such as Subversion. Entry and exit in the project implicate both a division of participants and the need to erect new infrastructural support. The popularization of distributed version control systems such as GIT, Bazaar and Mercurial is changing this default (as discussed above in *Multiplicity and Social Coding*), and creating more situations where the autonomous development of code, and the possibility of its repeated collaborative merging are rendered more explicit. One could say that the future is one where the fork, a separated initiative, is the basic state, always awaiting its moment of reintegration.

27. Forks vs. Knives

When you face a simple task and have all the capability and know-how necessary to accomplish it by yourself, there is really no reason for you to collaborate. And that's OK.

But when achieving the goal is hard, and the tools you have are insufficient, there is room for collaboration. In some cases collective action will be a part of the task itself—meaning the mobilization of group collaboration is a part of the task. Often this mobilization is a by product of the principal objective.

We defined intentionality and coordination of contributions as key to collaboration. But both intention and coordination actually raise the cost of collaboration, and in some cases makes it not worth the trouble.

> "Sometimes developers simply want to publish and share their work, not start a social movement. Sometimes they want to contribute to a project without going through masonic hazing rituals."

This quote from the Multiplicity and Social Coding chapter refers to Distributed Version Control Systems. The loose coordination enabled by these systems attempts to lower the cost of collaboration. By using these distributed collaborative tools the overhead inherent in establishing intention and coordination is reduced. In fact these system allow for a completely individualized practice. A Git user can work alone for years. By publishing her files online under a Free Software license she opens the door to a potential reappropriation or even future collaboration. She does not have to commit to contribute, she does not have to coordinate with anyone. She is not collaborating (yet).

This approach is similar to the principles advocated by the Free Culture movement. Share first and maybe collaborate later, or have others use your work, be it in an individual or a collaborative manner. But while distributed sharing platforms are common, DVCS excels in constantly switching between a coordinated action and an individuated one.

At any point of the development process Alice, a Git user, can inspect Bob's code repository and choose to fork (essentially duplicate) his code base to work on it separately. No permission is required, no coordination needed. At any point, Bob can pull Alice's changes and merge them back into his own repository.

This might seem trivial, but it's not. Centralized version control systems can make it technically easy to fork but are usually not sophisticated enough to make merging back easy. This has turned forking into a highly contested practice, as forking the code meant forking the project and dividing the community. DVCS makes forking and merging trivial and lowers the cost of collaboration.

But whilst distributing and individuating the process minimizes the need for intent and coordination, it may result in deemphasizing the collaborative act. By ensuring that 'you don't have to start a social movement', does it divorce itself from the social ideals of collaboration celebrated by many Free Software activists?

We argue it does not. You can still start a social movement if you like. You might actually have better tools to do so too, as the distributed process allow a larger autonomy for the individual members and less friction in governance and control.

*Glossary: **Architecture***

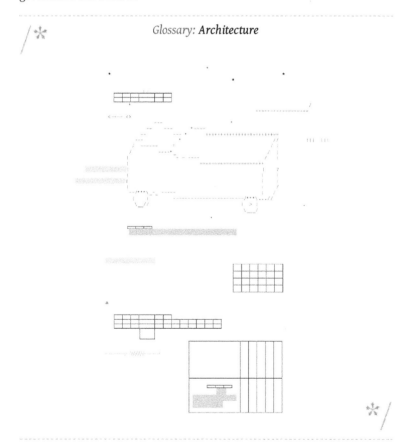

From Splicing Code to Assembling Social Solidarities

We can already see early examples of these approaches outside of Free Software. One of them is the Twitter Vote Report used in the 2008 US presidential elections <*twittervotereport.com*> and its later incarnation as SwiftRiver, a tool for crowdsourcing situational awareness:

> "Swift hopes to expand [Twitter Vote Report's] approach into a general purpose toolkit for crowdsourcing the semantic structuring of data so that it can be reused in other applications and visualizations. The developers of Swift are particularly interested in crisis reporting (Ushahidi) and international media criticism (Meedan), but by providing a general purpose crowdsourcing tool we hope to create a tool reusable in many contexts. Swift engages self-interested teams of "citizen editors" who curate publicly available information about a crisis or any event or region as it happens"
> —Swift Project, 2010 <*http://github.com/unthinkingly/swiftriver_rails*>

These activist hacker initiatives are realizing the potential of loosely coordinated distributed action. Its political power is entangled with its pragmatism, allowing the collaboration to fluidly shift between individual and collective action. In January 2010, as the horrors of the Haiti earthquake were unraveling, hackers around the world were mobilizing in unconference-style "CrisisCamps". These hackathons gathered individuals in physical space to "create technological tools and resources for responders to use in mitigating disasters and crises around the world."
<*crisiscommons.org/about-us*>

Contesting the Shock Doctrines

This type of external interest and action was previously reserved to human rights organization, media companies, governments and multinational corporations—all organizations that work in a pretty hierarchical and centralized manner. Now we see a new model emerge—a distributed networked collaboration of interested individuals contributing digital labor, not just money.

The political vacuum presented by these natural or man made crises leave room for a strong active force that often enforces a new political and economic reality. In her book titled The Shock Doctrine, author Naomi Klein describes how governments and businesses have exploited instances of political and economic instabilities in recent decades to dictate a neo-liberal agenda. In each case the interested powers were the first on the scene, imposing rigid rules of engagement and coordination, and justifying enforcement by the need to restore order.

In contrast, the activists are providing the tools and the know how for data production and aggregation. They are then actively assembling them into actionable datasets:

> "People on the ground need information, desperately. They need to know which symbols indicate that a house has already been searched, where the next food/water/medicine drop will be, and that the biscuits are good, and not expired. They also need entertainment, and news—à la Good Morning Vietnam. And messages of consolation, emotional support, solidarity, and even song and laughter. "
>
> —Jonnah Bossowitch

The model of individual autonomy and free association that enables the hackers coding is embedded into the assistance they propose, empowering communities on the ground. One of the hackers from the NYC CrisisCamp jokingly declares: *"Two sides get to play the shock doctrine game"*. It is obviously a drop in the ocean in comparison to the scale of the disaster and the years it would take to heal. Nations and corporations have long term interests and the resources that will probably keep them in the picture long after the networked effort will evaporate.

These are brave yet very early experiments in new political association. They are widely informed by experiments in collaboration and control in information economies. Most of them will not automatically translate to meat space. Especially at times of natural disaster, when food and medicine shortage occupies much of the human rights debate.

Open Leaders Failing Forward

The abundance of information technologies have also lowered the price of failure and made it an inherent part of the process. It's not about whether you will fail, it's about how you will fail. What will you learn from failure? How will you do things differently next time. This is what some call "failing forward".

The lowering costs of failure in distributed networked production allows for a more open emergence of leadership. One may provide leadership for a while and then get stuck, lose the lead, and be replaced by another one forking and leading in a different direction. This algorithmic logic justifies open access to knowledge and distribution of power that favors merit over entitlement. This is not a democracy, but a meritocracy. A meritocracy that favors technical expertise, free time, persistence and social skills. All traits that are definitely not evenly distributed.

Initiatives like FLOSS Manuals have acknowledged the importance of documentation for the collaborative process. To take real advantage of the network effect, we should learn to document failure, not only success.

In the past, experiments in alternative social organizations were hampered by limitations on the resources available within individual projects, and isolated by the costs of communication and coordination with kindred efforts. This was the case of the Cooperative movement, communes, the occupied factories in Argentina and other similar alternative social experiments. If we extend the notion of failing forward beyond the production of information, future results might look different.

New models of collaboration will continue to inform and alter our social relations. These political experiments are free for us to assess, free for us to fork and to try something different. Then, in the future after more development is done, and the commits have been tested, we will also be free to merge them back.

28. The Tyranny of Structurelessness

In the early 1970s, when feminism was just gaining steam, a woman named Jo Freeman wrote an essay that has since become a classic: "The Tyranny of Structurelessness." Though Freeman's essay was a response to the informal nature of women's "consciousness raising" groups popular during that period, it's also worth noting that second wave feminism emerged partly in reaction to the implicit and oppressive misogyny of the New Left, which united around an idealistic vision of decentralized "participatory democracy."

We've chosen to include an excerpt of this groundbreaking essay for various reasons. First of all, it underscores the vital role the women's movement had in theorizing, developing, and promoting non-hierarchical models of social justice organizing, an innovation they rarely get credited for. At the same time, however, it articulates the limits of these methods from an explicitly feminist perspective. Freeman's point, radically simplified, is that the disavowal of power too often masks its covert manipulation; informal elites can be more pernicious than formal ones because they deny their own existence.

It's no secret that software and technology industries are dominated by men, nor that many of the most visible figures writing about and promoting networked collaboration are male. We want to remind them that privilege is the product of complex social forces that cannot simply be wished away, no matter how loudly or frequently the word "open" is invoked. Those who like to believe that "on the Internet, no one knows you're a dog" are probably men (anyone remember that University of Maryland study reporting that chatters with female usernames got 25 times the number of malicious messages their masculine counterparts received?). The point is that offline prejudice, like offline privilege, carries over to our online relationships. The worry is that structurelessness can create a vacuum in which these imbalances and biases flourish.

> Contrary to what we would like to believe, there is no such thing as a structureless group. Any group of people of whatever nature that comes together for any length of time for any purpose will inevitably structure itself in some fashion. The structure may be flexible; it may vary over time; it may evenly or unevenly distribute tasks, power and resources over the members of the group. But it will be formed regardless of the abilities, personalities, or intentions of the people involved. The very fact that we are individuals, with different talents, predispositions, and backgrounds makes this inevitable. Only if we refused to relate or interact on any basis whatsoever could we approximate structurelessness—and that is not the nature of a human group.

This means that to strive for a structureless group is as useful, and as deceptive, as to aim at an "objective" news story, "value-free" social science, or a "free" economy. A "laissez faire" group is about as realistic as a "laissez faire" society; the idea becomes a smokescreen for the strong or the lucky to establish unquestioned hegemony over others. This hegemony can be so easily established because the idea of "structurelessness" does not prevent the formation of informal structures, only formal ones. Similarly "laissez faire" philosophy did not prevent the economically powerful from establishing control over wages, prices, and distribution of goods; it only prevented the government from doing so. Thus structurelessness becomes a way of masking power, and within the women's movement is usually most strongly advocated by those who are the most powerful (whether they are conscious of their power or not). As long as the structure of the group is informal, the rules of how decisions are made are known only to a few and awareness of power is limited to those who know the rules. Those who do not know the rules and are not chosen for initiation must remain in confusion, or suffer from paranoid delusions that something is happening of which they are not quite aware.

For everyone to have the opportunity to be involved in a given group and to participate in its activities the structure must be explicit, not implicit. The rules of decision-making must be open and available to everyone, and this can happen only if they are formalized. This is not to say that formalization of a structure of a group will destroy the informal structure. It usually doesn't. But it does hinder the informal structure from having predominant control and make available some means of attacking it if the people involved are not at least responsible to the needs of the group at large.

…

When informal elites are combined with a myth of "structurelessness," there can be no attempt to put limits on the use of power. It becomes capricious.

—Jo Freeman, "The Tyranny of Structurelessness," 1970

Glossary: **The Glossary of Tyranny**

A deviate of a neutral glossary. See also Glossary.

Glossary: **Tyranny**

Among us.

---

29. Free vs. Gratis Labor

What makes a collaboration "open"? Open appears to be an assertion of —though it is more accurately an aspiration towards—egalitarianism, inclusion, non-coerciveness, freedom.

But what kind of freedom? Free as in unscripted and improvisatory, free as in freely chosen, free as in unpaid, or free as in it won't tie you down?

As books like Jeff Howe's *Crowdsourcing: Why The Power of the Crowd is Driving the Future of Business* show, corporate America is ready to collaborate. They want to have an open relationship with their workforce, because who can beat free? And by turning their consumers into collaborators, the bond between the company and their customers is made even stronger. Meanwhile, everyday people are happy to help. Why? Howe says people do it for fun and for the "cred", otherwise known as the "emerging reputation economy."

In her oft-referenced essay "Free Labor: Producing Culture for the Digital Economy," Tiziana Terranova discusses free labor's complex relationship to capitalism.

> Free labor is a desire of labor immanent to late capitalism, and late capitalism is the field that both sustains free labor and exhausts it. It exhausts it by subtracting selectively but widely the means through which that labor can reproduce itself: from the burnout syndromes of Internet start-ups to underretribution and exploitation in the cultural economy at large. Late capitalism does not appropriate anything: it nurtures, exploits, and exhausts its labor force and its cultural and affective production. In this sense, it is technically impossible to separate neatly the digital economy of the Net from the larger network economy of late capitalism. Especially since 1994, the Internet is always and simultaneously a gift economy and an advanced capitalist economy. The mistake of the neoliberalists (as exemplified by the Wired group), is to mistake this coexistence for a benign, unproblematic equivalence.

As Terranova shows, we cannot pretend the gift economy of free (unpaid) collaboration exists totally apart from, untainted by, neoliberalism. The two are deeply interconnected, and becoming more so. Perhaps this is what our collaborative future really looks like.

Power exists after decentralization and neoliberalism operates in a distributed fashion—by smashing institutions (healthcare, schools, social welfare) and setting workers "free". Freedom, crowdsource, outsource, choice, autonomy, independence, innovation. We should not forget that these are also words central to the vocabulary of the market.

Free as in free speech? Free as in free beer? Free as in free world?
Free as in free markets? Free as in free labor? Using the word
"free" is ambiguous as it can be used to celebrate the excesses of
neoliberal capitalism and/or imply a confrontational position to this
regime. It is important to clarify what we mean when we say 'free'.
Who is free and what exactly are they free/freed from? Free is not
always good and structure is not always bad (see *The Tyranny of
Structurelessness*). Freedom and structure are not necessarily
oppositional. As Deleuze and Guattari caution, "Never believe that a
smooth space will suffice to save us." Freedom might not be such a
great thing if it means, like anonymous (see first section in the
book), that collective action produces terrorism, renunciation of
social responsibility and intolerance or, like free markets, that
openness produces extreme social and economic inequality.

Perhaps it is important to rethink our relation to freedom.

In this new world we are no longer bound to jobs that are required by law to stay with us, to support us; instead we're free to reinvent ourselves, to flit from project to project. This is the freedom of not being tied down, though the downside is that we're on our own. It's an open relationship, after all. We are freer, in many respects, but also replaceable. The crowd can take over; the crowd is cheaper, more efficient, less demanding.

These days, as the precarity of populations increases, we should ask ourselves, who has the free time to collaborate? Time is not an evenly distributed resource, as women working the "second shift" know too well. Who is able to work for "reputation," a substance invoked not just by Howe but also in the pages of this book? Only those who have enough to cover the basics.

30. Other People's Computers

Much of what we call collaboration occurs on web sites generally running software services. This is particularly true of collaboration among many distributed users. Direct support for collaboration, and more broadly for social features, is simply easier in a centralized context. It is possible to imagine a decentralized Wikipedia or Facebook, but building such services with sufficient ease of use, features, and robustness to challenge centralized web sites is a very difficult task.

Why does this matter? The web is great for collaboration, let's celebrate that! However, making it relatively easy for people to work together in the specific way offered by a web site owner is a rather impoverished vision of what the web and digital networks could enable, just as merely allowing people to run programs on their computers in the way program authors intended is an impoverished vision of personal computing.

Free software allows users control their own computing and to help other users by retaining the ability to run, modify, and share software for any purpose. Whether the value of this autonomy is primarily ethical, as often framed by advocates of the term free software, or primarily practical, as often framed by advocates of the term open source, any threat to these freedoms has to be of deep concern to anyone interested in the future of collaboration, both in terms of what collaborations are possible and what interests control and benefit from those collaborations. Kragen Sitaker frames the problem with these threats to freedom:

> "Web sites and special-purpose hardware [...] do not give me the same freedoms general-purpose computers do. If the trend were to continue to the extent the pundits project, more and more of what I do today with my computer will be done by special-purpose things and remote servers.
>
> What does freedom of software mean in such an environment? Surely it's not wrong to run a Web site without offering my software and databases for download. (Even if it were, it might not be feasible for most people to download them. IBM's patent server has a many-terabyte database behind it.)

I believe that software—open-source software, in particular—has the potential to give individuals significantly more control over their own lives, because it consists of ideas, not people, places, or things. The trend toward special-purpose devices and remote servers could reverse that.

—Kragen Sitaker, "people, places, things, and ideas ",
<*lists.canonical.org/pipermail/kragen-tol/1999-January/000322.html*>

What are the prospects and strategies for keeping the benefits of free software in an age of collaboration mediated by software services? One strategy, argued for in "The equivalent of free software for online services" by Kragen Sitaker (see <*lists.canonical.org/pipermail/kragen-tol/2006-July/000818.html*>), is that centralized services need to be re-implemented as peer-to-peer services that can run on computers as free software under users' control. This is an extremely interesting strategy, but a very long term one, for it is both a computer science challenge and a social one.

Abstinence from software services may be a naive and losing strategy in both the short and long term. Instead, we can both work on decentralization as well as attempt to build services that respect user's autonomy:

"Going places I don't individually control—restaurants, museums, retail stores, public parks—enriches my life immeasurably. A definition of "freedom" where I couldn't leave my own house because it was the only space I had absolute control over would not feel very free to me at all. At the same time, I think there are some places I just don't want to go—my freedom and physical well-being wouldn't be protected or respected there.

Similarly, I think that using network services makes my computing life fuller and more satisfying. I can do more things and be a more effective person by spring-boarding off the software on other peoples' computers than just with my own. I may not control your email server, but I enjoy sending you email, and I think it makes both of our lives better.

And I think that just as we can define a level of personal autonomy that we expect in places that belong to other people or groups, we should be able to define a level of autonomy that we can expect when using software on other people's computers. Can we make working on network services more like visiting a friends' house than like being locked in a jail?

We've made a balance between the absolute don't-use-other-people's-computers argument and the maybe-it's-OK-sometimes argument in the Franklin Street Statement. Time will tell whether we can craft a culture around Free Network Services that is respectful of users' autonomy, such that we can use other computers with some measure of confidence."

—Evan Prodromou, "RMS on Cloud Computing: "Stupidity"", CC BY-SA,
<*autonomo.us/2008/09/rms-on-cloud-computing-stupidity/*>

The Franklin Street Statement on Freedom and Network Services is an initial attempt to distill actions that users, service providers (the "other people" here), and developers should take to retain the benefits of free software in an era of software services:

> "The current generation of **network services** or **Software as a Service** can provide advantages over traditional, locally installed software in ease of deployment, collaboration, and data aggregation. Many users have begun to rely on such services in preference to software provisioned by themselves or their organizations. This move toward centralization has powerful effects on software freedom and user autonomy.

On March 16, 2008, a working group convened at the Free Software Foundation to discuss issues of freedom for users given the rise of network services. We considered a number of issues, among them what impacts these services have on user freedom, and how implementers of network services can help or harm users. We believe this will be an ongoing conversation, potentially spanning many years. Our hope is that free software and open source communities will embrace and adopt these values when thinking about user freedom and network services. We hope to work with organizations including the FSF to provide moral and technical leadership on this issue.

We consider network services that are **Free Software** and which share **Free Data** as a good starting-point for ensuring users' freedom. Although we have not yet formally defined what might constitute a 'Free Service', we do have suggestions that developers, service providers, and users should consider:

Developers of network service software are encouraged to:

- Use the GNU Affero GPL, a license designed specifically for network service software, to ensure that users of services have the ability to examine the source or implement their own service.
- Develop freely-licensed alternatives to existing popular but non-Free network services.
- Develop software that can replace centralized services and data storage with distributed software and data deployment, giving control back to users.

Service providers are encouraged to:

- Choose Free Software for their service.
- Release customizations to their software under a Free Software license.
- Make data and works of authorship available to their service's users under legal terms and in formats that enable the users to move and use their data outside of the service. This means:
  - Users should control their private data.
  - Data available to all users of the service should be available

under terms approved for Free Cultural Works or Open Knowledge.

Users are encouraged to:

- Consider carefully whether to use software on someone else's computer at all. Where it is possible, they should use Free Software equivalents that run on their own computer. Services may have substantial benefits, but they represent a loss of control for users and introduce several problems of freedom.
- When deciding whether to use a network service, look for services that follow the guidelines listed above, so that, when necessary, they still have the freedom to modify or replicate the service without losing their own data."

—Franklin Street Statement on Freedom and Network Services, CC BY-SA, <*autonomo.us/2008/07/franklin-street-statement/*>

As challenging as the Franklin Street Statement appears, additional issues must be addressed for maximum autonomy, including portable identifiers:

"A Free Software Definition for the next decade should focus on the user's overall autonomy- their ability not just to use and modify a particular piece of software, but their ability to bring their data and identity with them to new, modified software.

Such a definition would need to contain something like the following minimal principles:

1. data should be available to the users who created it without legal restrictions or technological difficulty.
2. any data tied to a particular user should be available to that user without technological difficulty, and available for redistribution under legal terms no more restrictive than the original terms.
3. source code which can meaningfully manipulate the data provided under 1 and 2 should be freely available.
4. if the service provider intends to cease providing data in a manner compliant with the first three terms, they should notify the user of this intent and provide a mechanism for users to obtain the data.
5. a user's identity should be transparent; that is, where the software exposes a user's identity to other users, the software should allow forwarding to new or replacement identities hosted by other software."

—Luis Villia, "Voting With Your Feet and Other Freedoms", CC BY-SA, <*tieguy.org/blog/2007/12/06/voting-with-your-feet-and-other-freedoms/*>

Fortunately the oldest, and at least until recently, the most ubiquitous network service—email—accommodates portable identifiers. (Not to mention that email is the lowest common denominator for much collaboration—sending attachments back and forth.) Users of a centralized email service like Gmail *can* retain a great deal of autonomy *if* they use an email address at a domain they control and merely route delivery to the service—though of course most users use the centralized provider's domain.

It is worth noting that the more recent and widely used, if not ubiquitous, instant messaging protocol XMPP as well as the brand new and little used Wave protocol have an architecture similar to email, though use of non-provider domains seems even less common, and in the case of Wave, Google is currently the only service provider.

It may be valuable to assess software services from the respect of community autonomy as well as user autonomy. The former may explicitly note requirements for the product of collaboration—non-private data, roughly—as well as service governance:

> In cases where one accepts a centralized web application, should one demand that application be somehow constitutionally open? Some possible criteria:
>
> - All source code for the running service should be published under an open source license and developer source control available for public viewing.
> - All private data available for on-demand export in standard formats.
> - All collaboratively created data available under an open license (e.g., one from Creative Commons), again in standard formats.
> - In some cases, I am not sure how rare, the final mission of the organization running the service should be to provide the service rather than to make a financial profit, i.e., beholden to users and volunteers, not investors and employees. Maybe. Would I be less sanguine about the long term prospects of Wikipedia if it were for-profit? I don't know of evidence for or against this feeling.
>
> —Mike Linksvayer, "Constitutionally open services", CC0,
> <*gondwanaland.com/mlog/2006/07/06/constitutionally-open-services/*>

Software services are rapidly developing and subjected to much hype, often referred to the buzzword Cloud Computing. However, some of the most potent means of encouraging autonomy may be relatively boring—for example, making it easier to maintain one's own computer and deploy slightly customized software in a secure and foolproof fashion. Any such development helps traditional users of free software as well as makes doing computing on one's own computer (which may be a "personal server" or virtual machine that one controls) more attractive.

Perhaps one of the most hopeful trends is relatively widespread deployment by end users of free software web applications like WordPress and MediaWiki. StatusNet, free software for microblogging, is attempting to replicate this adoption success. StatusNet also includes technical support for a form of decentralization (remote subscription) and a legal requirement for service providers to release modifications as free software via the AGPL.

This section barely scratches the surface of the technical and social issues raised by the convergence of so much of our computing, in particular computing that facilitates collaboration, to servers controlled by "other people", especially when these "other people" are a small number of large service corporations. The challenges of creating autonomy-respecting alternatives should not be understated.

One of those challenges is only indirectly technical: decentralization can make community formation more difficult. To the extent the collaboration we are interested in requires community, this is a challenge. However, easily formed but inauthentic and controlled community also will not produce the kind of collaboration we are interested in.

We should not limit our imagination to the collaboration facilated by the likes of Facebook, Flickr, Google Docs, Twitter, or other "Web 2.0" services. These are impressive, but then so was AOL two decades ago. We should not accept a future of collaboration mediated by centralized giants now, any more than we should have been, with hindsight, happy to accept information services dominated by AOL and its near peers.

Wikipedia is both held up as an exemplar of collaboration and is a free-as-in-freedom service: both the code and the content of the service are accessible under free terms. It is also a huge example of community governance in many respects. And it is undeniably a category-exploding success: vastly bigger and useful in many more ways than any previous encyclopedia. Other software and services enabling autonomous collaboration should set their sights no lower—not to merely replace an old category, but to explode it.

However, Wikipedia (and its MediaWiki software) are not the end of the story. Merely using MediaWiki for a new project, while appropriate in many cases, is not magic pixie dust for enabling collaboration. Affordances for collaboration need to be built into many different types of software and services. Following Wikipedia's lead in autonomy is a good idea, but many experiments should be encouraged in every other respect. One example could be the young and relatively domain-specific collaboration software that this book is being written with, Booki.

Software services have made "installation" of new software as simple as visiting a web page, social features a click, and provide an easy ladder of adoption for mass collaboration. They also threaten autonomy at the individual and community level. While there are daunting challenges, meeting them means achieving "world domination" for freedom in the most important means of production—computer-mediated collaboration —something the free software movement failed to approach in the era of desktop office software.

31. Science 2.0

Let the future tell the truth and evaluate each one according to his work and accomplishments. The present is theirs; the future, for which I really worked, is mine.

—Nikola Tesla

Science is a prototypical example of collaboration, from closely coupled collaboration within a lab to the very loosely coupled collaboration of the grant scientific enterprise over centuries. However, Science has been slow to adopt modern tools and methods for collaboration. And the efforts to adopt or translate those new tools and methods have been broadly (and loosely) characterized as "Science 2.0" and "Open Science", very roughly corresponding to "Web 2.0" and "Open Source".

Why Science 2.0? Didn't we claim in the chapter 'A Brief History of Collaboration' that "Web 2.0 is bullshit" as the "version number" of the Web as it conveys the incorrect sense that progress is not incremental and a marketing-driven message to "upgrade"?

For these same reasons Science 2.0 is appropriate. In general science hasn't made effective use the Web—translating and adopting the best practices of open collaboration on the web would constitute an "upgrade" and such an upgrade should be encouraged rhetorically.

This is largely the case due to Science's current setting in giant, slow to change institutions. But institutions, when they finally do change, can force broad change, quickly, as a matter of policy. Another reason the "upgrade" connotation is appropriate.

Open Access (OA) publishing is the vanguard—an effort to remove a major barrier to distributed collaboration in science—the high price of journal articles, effectively limiting access to researchers affiliated with wealthy institutions. Access to Knowledge (A2K) emphasizes the equality and social justice aspects of opening access to the scientific literature.

The OA movement has met with substantial and increasing success recently. The Directory of Open Access Journals <*www.doaj.org*> lists 5.148 journals as of July 2010. The Public Library of Science's top journals are in the first tier of publications in their fields. Traditional publishers are investing in OA, such as Springer's acquisition of large OA publisher BioMed Central, or experimenting with OA, for example Nature Precedings.

In the longer term, OA may lead to improving the methods of scientific collaboration, like peer review, and allowing new forms of meta-collaboration. An early example of the former is PLoS ONE, a rethinking of the journal as an electronic publication without a limitation on the number of articles published and with the addition of user rating and commenting. An example of the latter would be machine analysis and indexing of journal articles, potentially allowing all scientific literature to be treated as a database, and therefore queryable—at least all OA literature. These more sophisticated applications of OA often require not just access, but permission to redistribute and manipulate, thus a rapid movement to publication under a Creative Commons license that permits any use with attribution—a practice followed by both PLoS and BioMed Central.

Scientists have also adopted web tools to enhance collaboration within a working group as well as to facilitate distributed collaboration. Wikis and blogs have been purposed as as open lab notebooks under the rubric of "Open Notebook Science". Connotea is a tagging platform (they call it "reference management") for scientists. These tools help "scale up" and direct the scientific conversation, as explained by Michael Nielsen:

> "You can think of blogs as a way of *scaling up* scientific conversation, so that conversations can become widely distributed in both time and space. Instead of just a few people listening as Terry Tao muses aloud in the hall or the seminar room about the Navier-Stokes equations, why not have a few thousand talented people listen in? Why not enable the most insightful to contribute their insights back?
>
> ...
>
> Stepping back, what tools like blogs, open notebooks and their descendants enable is filtered access to new sources of information, and to new conversation. The net result is a *restructuring of expert attention*. This is important because expert attention is the ultimate scarce resource in scientific research, and the more efficiently it can be allocated, the faster science can progress."
>
> —Michael Nielsen, "Doing science online", <*michaelnielsen.org/blog/doing-science-online/*>

OA and adoption of web tools are only the first steps toward utilizing digital networks for scientific collaboration. Science is increasingly computational and data-intensive: access to a completed journal article may not contribute much to allowing other researcher's to build upon one's work—that requires publication of all code and data used during the research used to produce the paper. Publishing the entire "research compendium" under appropriate terms (e.g. usually public domain for data, a free software license for software, and a liberal Creative Commons license for articles and other content) and in open formats has recently been called "reproducible research"—in computational fields, the publication of such a compendium gives other researches all of the tools they need to build upon one's work.

Standards are also very important for enabling scientific collaboration, and not just coarse standards like RSS. The Semantic Web and in particular ontologies have sometimes been ridiculed by consumer web developers, but they are necessary for science. How can one treat the world's scientific literature as a database if it isn't possible to identify, for example, a specific chemical or gene, and agree on a name for the chemical or gene in question that different programs can use interoperably? The biological sciences have taken a lead in implementation of semantic technologies, from ontology development and semantic databases to in-line web page annotation using RDFa.

Of course all of science, even most of science, isn't digital. Collaboration may require sharing of physical materials. But just as online stores make shopping easier, digital tools can make sharing of scientific materials easier. One example is the development of standardized Materials Transfer Agreements accompanied by web-based applications and metadata, potentially a vast improvement over the current choice between ad hoc sharing and highly bureaucratized distribution channels.

"Open Innovation" is a practice that is somewhere between open science and business. Open Innovation refers to a collection of tools and methods for enabling more collaboration. Some of these Open Innovation tools include crowdsourcing of research expertise which is being lead by a company called InnoCentive, patent pools, end-user innovation which Erik von Hippel documented in *Democratizing Innovation*, and wisdom of the crowds methods such as prediction markets.

Reputation is an important question for many forms of collaboration, but particularly in science, where careers are determined primarily by one narrow metric of reputation—publication. If the above phenomena are to reach their full potential, they will have to be aligned with scientific career incentives. This means new reputation systems that take into account, for example, re-use of published data and code, and the impact of granular online contributions, must be developed and adopted.

From the grand scientific enterprise to business enterprise modern collaboration tools hold great promise for increasing the rate of discovery, which sounds prosaic, but may be our best tool for solving our most vexing problems. John Wilbanks, Vice President for Science at Creative Commons often makes the point like this: "We don't have any idea how to solve cancer, so all we can do is increase the rate of discovery so as to increase the probability we'll make a breakthrough."

Science 2.0 also holds great promise for allowing the public to access current science, and even in some cases collaborate with professional researchers. The effort to apply modern collaboration tools to science may even increase the rate of discovery of innovations in collaboration!

32. Beyond Education

Education has a complicated history, including swings between decentralization -those loose associations of students and teachers typifying some early European universities such as Oxford- to centralized control by the state or church. It's easy to imagine that in some of these cases teachers had great freedom to collaborate with each other or that learning might be a collaboration among students and teacher, while in others, teachers would be told what to teach, and students would learn that, with little opportunity for collaboration.

The current and unprecedented wealth in the Global North has brought near universal literacy and enrollment in primary education, along with impressive research universities and increasing enrollment in university and graduate programs. This apparent success masks that we are in an age of unprecedented inequality, centralized control, driven by standards politically determined at the level of large jurisdictions, and a model in which teachers teach how to take tests and both students and teachers are consumers of educational materials created by large publishers. Current educational structures and practices do not take advantage of the possibilities offered by collaboration tools and methods and in some cases are in opposition to use of such tools.

Much as the disconnect between the technological ability to access and build upon and the political and economic reality of closed access in scientific publishing created the Open Access (OA) movement, the disconnect between what is possible and what is practiced in education has created collaborative responses.

Open Educational Resources

The Open Educational Resources (OER) movement encourages the availability of educational materials for free use and remix—including textbooks and also any materials that facilitate learning. As in the case of OA, there is a strong push for materials to be published under liberal Creative Commons licenses and in formats amenable to reuse in order to maximize opportunities for latent collaboration, and in some cases to form the legal and technical basis for collaboration among large institutions.

OpenCourseWare (OCW) is the best known example of a large institutional collaboration in this space. Begun at MIT, over 200 universities and associated institutions have OCW programs, publishing course content and in many cases translating and reusing material from other OCW programs.

Connexions, hosted by Rice University, is considered by many as an example of an OER platform facilitating large scale collaborative development and use of granular "course modules" which currently number over 15,000. The Connexions philosophy page is explicit about the role of collaboration in developing OER:

"Connexions is an environment for collaboratively developing, freely sharing, and rapidly publishing scholarly content on the Web. Our Content Commons contains educational materials for everyone—from children to college students to professionals—organized in small modules that are easily connected into larger collections or courses. All content is free to use and reuse under the Creative Commons "attribution" license.

Content should be modular and non-linear
Most textbooks are a mass of information in linear format: one topic follows after another. However, our brains are not linear—we learn by making connections between new concepts and things we already know. Connexions mimics this by breaking down content into smaller chunks, called modules, that can be linked together and arranged in different ways. This lets students see the relationships both within and between topics and helps demonstrate that knowledge is naturally interconnected, not isolated into separate classes or books.

Sharing is good
Why re-invent the wheel? When people share their knowledge, they can select from the best ideas to create the most effective learning materials. The knowledge in Connexions can be shared and built upon by all because it is reusable:

- **technologically:** we store content in XML, which ensures that it works on multiple computer platforms now and in the future.
- **legally:** the Creative Commons open-content licenses make it easy for authors to share their work—allowing others to use and reuse it legally—while still getting recognition and attribution for their efforts.
- **educationally:** we encourage authors to write each module to stand on its own so that others can easily use it in different courses and contexts. Connexions also allows instructors to *customize* content by overlaying their own set of links and annotations. Please take the Connexions Tour and see the many features in Connexions.

Collaboration is encouraged
Just as knowledge is interconnected, people don't live in a vacuum. Connexions promotes communication between content creators and provides various means of collaboration. Collaboration helps knowledge grow more quickly, advancing the possibilities for new ideas from which we all benefit."

However, one major question with Connexions that needs to be asked is whether it really is a collaborative platform or is it a platform for sharing content. Authors that make their content available as "modules" for others to remix into collections, for example, might not be considered a collaborative activity. Although popularly considered collaborative it probably is better seen as a successful mechanism for sharing individually authored OER materials. Discussions and connections between contributors might be made around this shared content which might turn into collaborations that occur outside of the platform but Connexions itself as platform for collaboration is mostly aspirational.

There is also P2PU <p2pu.org/> and Wikiversity <en.wikiversity.org> which are trying to establish remote and collaborative pegagogical strategies, and another interesting group producing OER materials is the Teaching Open Source group. <www.teachingopensource.org>.

Beyond the institution

OER is not only used in an institutional context—it is especially a boon for self-learning. OCW materials are useful for self-learners, but OCW programs generally do not actively facilitate collaboration with self-learners. A platform like Connexions is more amenable to such collaboration, while wiki-based OER platforms have an even lower barrier to contribution that enable self-learners (and of course teachers and students in more traditional settings) to collaborate directly on the platform. Wiki-based OER platforms such as Wikiversity and WikiEducator make it even easier for learners and teachers in all settings to participate in the development and repurposing of educational materials.

Self-learning only goes so far. Why not apply the lessons of collaboration directly to the learning process, helping self-learners help each other? This is what a project called Peer 2 Peer University has set out to do:

> "The mission of P2PU is to leverage the power of the Internet and social software to enable communities of people to support learning for each other. P2PU combines open educational resources, structured courses, and recognition of knowledge/learning in order to offer high-quality low-cost education opportunities. It is run and governed by volunteers."

Scaling educational collaboration

As in the case of science, delivering the full impact of the possibilities of modern collaboration tools requires more than simply using the tools to create more resources. For the widest adoption, collaboratively created and curated materials must meet state-mandated standards and include accompanying assessment mechanisms.

> *Glossary: Educational Intervention*
>
> The book sprint might be applied to PhD dissertations. The book sprint could be extended to a thesis written over two weeks by 20 people. Is there a venture capitalist out there who would like to found "SprintMyDissertation.com"?

While educational policy changes may be required, perhaps the best way for open education communities to convince policymakers to make these changes is to develop and adopt even more sophisticated collaboration tools, for example reputation systems for collaborators and quality metrics, collaborative filtering and other discovery mechanisms for educational materials. One example are "lenses" at Connexions <cnx.org/lenses>, which allow one to browse resources specifically endorsed by an organization or individual that one trusts.

Again, similar to science, clearing the external barriers to adoption of collaboration may result in general breakthroughs in collaboration tools and methods.

33. How Would It Translate?

There is a movement spearheaded by Aspiration <*www.aspirationtech.org*> as a collaborative practice. This new practice is dubbed "open translation." There are abundant examples of community translation, however the tools required are primitive or simply don't exist and hence the opportunity for the practice to gather more momentum are stunted.

Ethan Zuckerman has commented on the need for a 'polygot Internet' and the need for collaborative translation:

> "The polyglot Internet demands that we explore the possibility and power of distributed human translation. Hundreds of millions of Internet users speak multiple languages; some percentage of these users are capable of translating between these. These users could be the backbone of a powerful, distributed peer production system able to tackle the audacious task of translating the Internet.
>
> We are at the very early stages of the emergence of a new model for translation of online content—"peer production" models of translation. Yochai Benkler uses the term "peer production" to describe new ways of organizing collaborative projects beyond such conventional arrangements as corporate firms. Individuals have a variety of motives for participation in translation projects, sometimes motivated by an explicit interest in building intercultural bridges, sometimes by fiscal reward or personal pride. In the same way that open source software is built by programmers fueled both by personal passion and by support from multinational corporations, we need a model for peer-produced translation that enables multiple actors and motivations.
>
> To translate the Internet, we need both tools and communities. Open source translation memories will allow translators to share work with collaborators around the world; translation marketplaces will let translators and readers find each other through a system like Mechanical Turk, enhanced with reputation metrics; browser tools will let readers seamlessly translate pages into the highest-quality version available and request future human translations. Making these tools useful requires building large, passionate communities committed to bridging a polyglot web, preserving smaller languages, and making tools and knowledge accessible to a global audience."

—Ethan Zuckerman, 2009
<*www.ethanzuckerman.com/blog/the-polyglot-internet/*>

The gaps in the tools and practices for collaborative translation have been been documented in the Open Translations Tools book <*en.flossmanuals.net/OpenTranslationTools*> which was the result of a Book Sprint coordinated by FLOSS Manuals and Aspiration. The content below comes from the chapter 'The Current State' which identifies the tools and processes required to catalyze this emergent field.

Workflow support

Though a number of 'Open Translation Tools' provide limited support for translation workflow processes, there is currently no tool or platform with rich and general support for managing and tracking a broad range of translation tasks and workflows. The internet has made possible a plethora of different collaborative models to support translation processes. But there are few FLOSS tools to manage those processes: tracking assets and state, role and assignments, progress and issues. While tools like Transifex provide support for specific workflows in specific communities, generalized translation workflow tools are still few in number. An ideal Open Translation tool would understand the range of roles played in translation projects, and provide appropriate features and views for users in each role. As of this writing, most Open Translation tools at best provide workflow support for the single type of user which that tool targets.

Distributed translation with memory aggregation

As translation and localization evolve to more online-centric models, there is still a dearth of tools which leverage the distributed nature of the Internet and offer remote translators the ability to contribute translations to sites of their choosing. As of this writing, Worldwide Lexicon is the most advanced platform in this regard, providing the ability for blogs and other open content sites to integrate distributed translation features into their interfaces. In addition, there needs to be a richer and more pervasive capture model for content translated through such distributed models, in order to aggregate comprehensive translation memories in a range of language pairs.

Interoperability

The lack of integration and interoperability between tools means both frustration for users and feature duplication by developers. Different communities have their own toolkits, but it is difficult for a translation project to make coherent use of a complete tool set. Among the interoperability issues which require further attention in the Open Translation tools ecology:

- Common programming interfaces for tools to connect, share data and requests, and collect translation memories and other valuable data.
- Plugins for content management systems to export content into *PO-files* (a standardized file format for storing translated phrases), so that content can be translated by the wealth of tools that offer PO support.
- Better integration between different projects, including shared glossaries, common user interfaces and subsystems, and rich file import/export.
- Generic code libraries for common feature requirements. "gettext" stands out as one of the most ubiquitous programming interfaces in the Open Translation arena, but many more interfaces and services could be defined and adopted to maximize interoperability of both code and data.

Review Processes

Tools for content review are also lacking; features for quality review should be focused on distributed process and community-based translation. As such reviews can be a delicate matter, the ideal communication model when there are quality problems is to contact the translator, but timing can be an issue. In systems with live posts and rapid translation turnaround, quick review is important and it may not be possible to reconnect with the content translator in a timely fashion.

34. Death is not the end

> *"There is nothing in the world to which every man has a more unassailable title than to his own life and person."*
> —Schopenhauer, *On Suicide*

Last year, the online world was surprised by two applications. They didn't offer faster, deeper or richer ways to engage, update or collaborate with other people. Quite the contrary! They encouraged users to liberate themselves from their needy, over-consuming virtual identities and jump back to the world of flesh meetings, slow readings and the realities of unpokeability by committing ritual online suicide.

> As the Seppuku restores samurai's honor as a warrior, in the same way, Seppukoo.com deals with the liberation of the digital body from any identity constriction in order to help people discover what happens after their virtual life and to rediscover the importance of being anyone, instead of pretending to be someone. [Seppukoo.com]

> This machine lets you delete all your energy sucking social-networking profiles, kill your fake virtual friends, and completely do away with your Web2.0 alterego. [Web 2.0 Suicide Machine]

The artist behind Seppukoo offered an artisan service; if you gave him your name and password, he would do the job for you, but The hackers behind the Web 2.0 Suicide Machine automatized it, turning a philosophical joke into a tool of insurrection. After more than a thousand ritual suicides, both sites received a cease and desist letter from Facebook's legal adviser, who accused them of asking other users to share their login data, entering other people's accounts, collecting other user's information, spamming and using Facebook's Intellectual Property without permission.

Seppukoo.com terminated its activities. The Web 2.0 Suicide Machine didn't, and it was consequently blocked from accessing Facebook accounts, along with all the project team members. That didn't stop them for long; the case had already made it to the newspapers and, while the service was often unavailable, it was not due to Facebook but by the waves of suicides hitting their server.

Glossary: **Mythologies**

Do we diminish, trying to erase, suppress, repress the mythological in our cultural environment? Does it not re-appear in myths of hacking, sharing, collaborating? What is mythological about these practices?

Curiously enough, all but one of these accusations refer to Facebook's own Statements of Rights and Responsibility. The Copyright infringement could possibly hold water, though it could also be protected under the Fair use for parody, "the use of some elements of a prior author's composition to create a new one that, at least in part, comments on that author's works." But Facebook claimed both websites were breaching the end user agreement, *and they were no user*. If someone had breached that contract, it was the Facebook suicides who "shared your password, let anyone else access your account, or do anything else that might jeopardize the security of your account". Why were the websites held responsible for other people's 2.0 crimes?

Fight for your right to die

"When you deactivate your account, no one can see your profile, but your information is saved in case you decide to reactivate later," the company told the newspaper. As it turns out, suicide is not a crime in Facebook, as it is in most western countries. The company would be quite reluctant to sue their ex-users for terminating their relationship with them, but it is entitled to make it difficult. "Users rely on us to protect their data and enforce the privacy decisions they make on Facebook -their spokesman insists.- We take this trust seriously and work aggressively to protect it". Even against their will.

The very real crime committed by the Web 2.0 Suicide Machine is facilitating and encouraging the means for suicide, an option that many users might have never thought about and that would take them quite a while to accomplish on their own.

With their help, dying can be like this: you choose the community you want to leave -MySpace, Facebook, Linkedin, Twitter-, give them your user name and login password, and everything in it, friends, connections, tweets, favorite, photos, will disappear. The only remaining sign of your 2.0 existence will be an empty profile with no more data than your last words: your grave. And there is no way back.

"Seamless connectivity and rich social experience offered by web 2.0 companies are the very antithesis of human freedom", explained the Suicide Machine spokesman in an interview to the BBC. A life owned by a dot.com company is not worth living, specially at the cost of the real one. Facebook friends are not real friends, and Facebook suicide is not really dying, but if we put together the hours dedicate to befriending, connecting, introducing, banning, integrating, supporting, comparing and managing their social lives, some would agree that virtual existence is rendering the real one empty, without giving much back.

Disconnecting people

The right to life is an inalienable right inherent in us by virtue of our existence, but the right to live in a server owned and regulated by an online corporation is a different thing. Why sacrifice one for the other? In getafirstlife.com (2006), artist Darren Barefoot argued for living IRL: "Go Outside. Membership is Free". Sepukoo and the Web 2.0 Suicide Machine have reversed the 2.0 obsession of belonging back to the right and the need to be one's own, "disconnecting people from each other and transforming the individual suicide experience into an exciting social experience".

The ritual itself is essential, as it stands as a gesture of independence from the platform, the community and the commercial interests of the big company behind it. Historically, user content based websites have been naturally reluctant to let their users go. Disappointed users got used to abandoning their virtual egos in a limbo of non-updated ghost user accounts. Limitations on Removal and clauses about uncontrollable and eternal Backup servers are common part of the EULA terms. Ghost status is not enough, as the suicide assistants remind us. You have to pull yourself away from the network entirely. Like all good collaborations you have to have the right to walk away and leave nothing behind.

Futures

35. Free as in Free World

Freedom of information and the various freedoms that attend it—the freedom to share, to know, to hack, to fork, to modify—reach a dead end in things that can't be copied without someone having less of the original (the problem of rivalrous vs. non-rivalrous goods in economic jargon). If there's a wish woven through these pages it's that non-rivalrous sharing will somehow be extended to the material realm, that spreading information and collaborating through a network will sow the seeds of a new culture less fixated on ownership, more prone to cooperation.

The problem is, it's one thing to pass on a file, while retaining a perfect copy for yourself, and quite another to fairly allocate valuable and essential finite material resources like land and water. Let's face it: human beings even hoard immaterial, intangible resources (we're thinking of things like power, privilege, and authority). Given that this is the case, how can we hope to make a leap from networked collaboration towards greater social equity? Is the type of collaboration we're talking about here even a first step, or is it a distraction? How, we wonder, can things like free software, free culture, and p2p be leveraged to encourage a more equal distribution of resources, a more even distribution of power, a dispersal of knowledge and influence?

One way to use this book is as a guide—sometimes critical—to signs on the horizon that may point to more positive collaborative futures. There is much more to be said—see *Things We Ended Up Not Including*—and even more to be done to make a collaborative, free, and positive future into reality.

36. Networked Solidarity

> "There is no guarantee that networked information technology will lead to the improvements in innovation, freedom, and justice that I suggest are possible. That is a choice we face as a society. The way we develop will, in significant measure, depend on choices we make in the next decade or so."
> —Yochai Benkler, *The Wealth of Networks: How Social Production Transforms Markets and Freedom*

Postnationalism

Catherine Frost, in her 2006 paper *Internet Galaxy Meets Postnational Constellation: Prospects for Political Solidarity After the Internet* evaluates the prospects for the emergence of postnational solidarities abetted by Internet communications leading to a change in the political order in which the responsibilities of the nation state are joined by other entities. Frost does not enumerate the possible entities, but surely they include supernational, transnational, international, and global in scope and many different forms, not limited to the familiar democratic and corporate.

The verdict? Characteristics such as anonymity, agnosticism to human fatalities and questionable potential for democratic engagement make it improbable that postnational solidarities with political salience will emerge from the Internet—anytime soon. However, Frost acknowledges that we could be looking in the wrong places, such as the dominant English-language Web. Marginalized groups could find the Internet a more compelling venue for creating new solidarities.

And this:

> "Yet we know that when things change in a digital age, they change fast. The future for political solidarity is not a simple thing to discern, but it will undoubtedly be an outcome of the practices and experiences we are now developing."

Could the collaboration mechanisms discussed in this book aid the formation of politically salient postnational solidarities? Significant usurpation of responsibilities of the nation state seems unlikely soon. Yet this does not bar the formation of communities that contest with the nation state for intensity of loyalty, in particular when their own collaboration is threatened by a nation state. As an example we can see global responses from free software developers and bloggers to software patents and censorship in single jurisdictions.

Trans[national]gression

If political solidarities could arise out of collaborative work and threats to it, then collaboration might alter the power relations of work. Both globally and between worker and employer—at least incrementally.

Workers are not permitted the freedom granted to traders and capitalists over the last half century, during which barriers to trade and investment were greatly reduced. People in jurisdictions with less opportunity are as locked into politically institutionalized underemployment and poverty as were non-whites in Apartheid South Africa, while the populations of wealthy jurisdiction are as privileged as whites were in the same regime, as explained by Yves Bonnardel and David Olivier's *Manifesto for the Abolition of International Apartheid,* *<webspace.utexas.edu/hcleaver/www/wk2abolition.html>*:

> The ethical and political principle of equality of all individuals of the human species is now acknowledged by nearly all. It is almost universally accepted that any discrimination between human individuals based on an arbitrary criterion is unjust and must be abolished.
>
> Since the end of interracial apartheid in South Africa, no longer any state in the world openly practices discrimination between humans based on the arbitrary criterion of skin color. Today, however, another equally arbitrary criterion is still accepted and applied by virtually every state in the world. For a human individual to have been born in some a particular place, from parents of some particular nationality, and thus to possess himself some particular nationality, is a matter of chance, and cannot be taken as a non-arbitrary criterion of discrimination.
>
> Following this arbitrary criterion of nationality, states either grant or deny human individuals the right to dwell on their territories as well as access to the social benefits that are granted to the natives. Just like interracial apartheid in South Africa, this arbitrary discrimination would be but a relatively harmless absurdity if its consequences were a mere separation. But the reality of the world we live in is marked by the existence of vast areas in which most inhabitants suffer from severe poverty and high rates of mortality; and of other areas in which inhabitants live in conditions that, though not always good, are for the least considerably better than the conditions that prevail in the poor areas. The refusal to allow certain individuals to live in rich countries on the basis of their nationality is *de facto*, just like interracial apartheid, an arbitrary denial of often vital benefits granted to others.

We therefore recognize as fundamentally contrary to the ethical and political principle of human equality the state laws and regulations, particularly those of the rich states, that deny individuals the right to enter and live on their territories, and access to social benefits, on the basis of their nationalities. We demand the abolition of this international apartheid, and demand that all appropriate measures be taken to render this abolition effective as quickly as possible.

As a consequence of the ethical and political principle of human equality, we recognize these laws and regulations as illegitimate. We demand that they be abolished, and that every human being, whatever eir nationality, be permitted to live on the territory of any state, and receive equally the social benefits that are granted to the natives.

We declare ourselves under no obligation to respect these illegitimate laws, and ready, should the case arise, to transgress them and to help others to transgress them.

What does this have to do with collaboration? This system of labor is immobilized by politically determined discrimination. It is not likely this system will change without the formation of new postnational orders. However, it is conceivable that as collaboration becomes more economically important—as an increasing share of wealth is created via distributed collaboration—the inequalities of the current system could be *mitigated*. That is simply because distributed collaboration does not require physical movement across borders.

Put more boldly, distributed collaboration is a means to transgress the system of International apartheid condemned by Bonnardel and Olivier. The effect and effectiveness of transgression is always hotly debated. However, it is also possible that open collaboration could alter relationships between some workers and employers in the workers' favor both in local and global markets.

Control of the means of production

Open collaboration changes which activities are more efficient inside or outside of a firm. Could the power of workers relative to firms also be altered?

"Intellectual property rights prevent mobility of employees in so far as their knowledge is locked in in a proprietary standard that is owned by the employer. This factor is all the more important since most of the tools that programmers are working with are available as cheap consumer goods (computers, etc.). The company holds no advantage over the worker in providing these facilities (in comparison to the blue-collar operator referred to above whose knowledge is bound to the Fordist machine park). When the source code is closed behind copyrights and patents, however, large sums of money is required to access the software tools. In this way, the owner/firm gains the edge back over the labourer/programmer.

This is were GPL comes in. The free license levels the playing field by ensuring that everyone has equal access to the source code. Or, putting it in Marxist-sounding terms, through free licenses the means of production are handed back to labour. [...] By publishing software under free licences, the individual hacker is not merely improving his own reputation and employment prospects, as has been pointed out by Lerner and Tirole. He also contributes in establishing a labour market where the rules of the game are completely different, for him and for everyone else in his trade. It remains to be seen if this translates into better working conditions,higher salaries and other benefits associated with trade unions. At least theoretically the case is strong that this is the case. I got the idea from reading Glyn Moody's study of the FOSS development model, where he states: "Because the 'product' is open source, and freely available, businesses must necessarily be based around a different kind of scarcity: the skills of the people who write and service that software." (Moody, 2001, p.248) In other words, when the source code is made available to everyone under the GPL, the only thing that remains scarce is the skills needed to employ the software tools productively. Hence, the programmer gets an edge over the employer when they are bargaining over salary and working conditions.

It bears to be stressed that my reasoning needs to be substantiated with empirical data. Comparative research between employed free software programmers and those who work with proprietary software is required. Such a comparison must not focus exclusively on monetary aspects. As important is the subjective side of programming, for instance that hackers report that they are having more fun when participating in free software projects than they work with proprietary software (Lakhani & Wolf, 2005). Neither do I believe that this is the only explanation to why hackers use GPL. No less important are the concerns about civil liberties and the anti-authoritarian ethos within the hacker subculture. In sum, hackers are a much too heterogeneous bunch for them all to be included under a single explanation. But I dare to say that the labour perspective deserves more attention than it has been given by popular and scholarly critics of intellectual property till now. Both hackers and academic writers tend to formulate their critique against intellectual property law from a consumer rights horizon and borrow arguments from a liberal, political tradition. There are, of course, noteworthy exceptions. People like Eben Moglen,

Slavoj Zizek and Richard Barbrook have reacted against the liberal ideology implicit in much talk about the Internet by courting the revolutionary rhetoric of the Second International instead. Their ideas are original and eye-catching and often full of insight. Nevertheless, their rhetoric sounds oddly out of place when applied to pragmatic hackers. Perhaps advocates of free sotftware would do better to look for a counter-weight to liberalism in the reformist branch of the labour movement, i.e. in trade unionism. The ideals of free software is congruent with the vision laid down in the "Technology Bill of Rights", written in 1981 by the International Association of Machinists:

"The new automation technologies and the sciences that underlie them are the product of a world-wide, centuries-long accumulation of knowledge. Accordingly, working people and their communities have a right to share in the decisions about, and the gains from, new technology" (Shaiken, 1986, p.272)."

—Johan Söderberg, *Hackers GNUnited!*, CC BY-SA, <*freebeer.fscons.org*>

Perhaps open collaboration can only be expected to slightly tip the balance of power between workers and employers and change measured wages and working conditions very little. However, it is conceivable, if fanciful, that control of the means of production could nurture a feeling of autonomy that empowers further action outside of the market.

Autonomous individuals and communities

Glossary: Autonomy

Autonomy is a concept found in moral, political, and bioethical philosophy. Within these contexts it refers to the capacity of a rational individual to make an informed, un-coerced decision. In moral and political philosophy, autonomy is often used as the basis for determining moral responsibility for one's actions. <*en.wikipedia.org/wiki/Autonomy*> The work of late twentieth-century thinkers and feminist scholars problematizes the notion that an individual subject could either precede all social formations or could possibly make rational decisions. Instead the body is seen as a site in which all manner of social forces are made manifest, articulated in physiological, psychological and biological ways. Body, mind, consciousness are sites of domination and subjection through modulation (Foucault). We enact power and power runs through us. Subjectivity is not an issue of an individual self but an agglomeration and enactment of social and political forces.

In short, do we always know whose will we are choosing? It is worthwhile to be suspicious of those people and projects who claim to be autonomous.

Free Software and related methodologies can give individuals autonomy in their technology environments. It might also give individuals a measure of additional autonomy in the market (or increased ability to stand outside it). This is how Free and Open Source Software is almost always characterized, when it is described in terms of freedom or autonomy—giving individual users freedom, or allowing organizations to not be held ransom to proprietary licenses.

However, communities that exist outside of the market and state obtain a much greater autonomy. These communities have no need for the freedoms discussed above, even if individual community members do. There have always been such communities, but they did not possess the ability to use open collaboration to produce wealth that significantly competes, even supplants, market production. This ability makes these autonomous organizations newly salient.

Furthermore, these autonomous communities (Debian and Wikipedia are the most obvious examples) are pushing new frontiers of governance necessary to scale their collaborative production. Knowledge gained in this process could inform and inspire other communities that could become reinvigorated and more effective through the implementation of open collaboration, including community governance. Such communities could even produce postnational solidarities, especially when attacked.

Do we know how to get from here to there? No. But only through experimentation will we find out. If a more collaborative future is possible, obtaining it depends on the choices we make today.

37. Free Culture in Cultures that are Not Free

> …it may not be the people with the most extensive access or highest profile online who will champion deep social and political change, if such is to come about. Instead, it is the groups with limited access, just enough to see what they are missing out on, who may have the most to gain from pioneering new modes of social relations, meaning, and engagement. Nor is it straightforward to suppose that such innovations will revolve around the Internet itself or other global structures. The innovations may well take contrasting forms, even as they take full advantage of the new capabilities and possibilities that the Internet introduces.
>
> Ironically, then, it may be the Internet's capacity to heighten the experience of exclusion, to promote awareness of a population's marginal and disenfranchised status, that represents its greatest potential for change.
>
> —Catherine Frost, *Internet Galaxy Meets Postnational Constellation: Prospects for Political Solidarity After the Internet*

The effect of the Internet on the Arab world is complex. While the Arab dictatorships are exercising extreme censorship and tight monitoring of online communication, the economic benefits of networking technology have had a relatively low impact on these countries. The Arab world has been a low priority for Western media corporations who were not interested in bridging the cultural divide. This divide is indeed complex, with the so-called "War on Terror" tainting foreign interests and with the culture of software piracy making this market even less economically attractive.

The Arabic language itself is posing a specially complex gap on its own. While localization of software has been relatively easy within most Latin languages, localization into Arabic, Farsi, Urdu and Hebrew requires bi-directional treatment of the text to account for the right-to-left directionality. This complicated mirroring problem together with the complex dynamics of the language amount for very high software localization costs.

"What the Web Can Be"

In 2003 a joint Israeli/Palestinian team attempted to pressure the Macromedia corporation to fully support right-to-left languages in its Flash plug-in. They released a petition under the banner "The Right to Flash" and contested Macromedia's motto "What the web can be" requiring to be included in the company's vision. The thousands of signatures and many blog posts published on the topic did nothing to budge the corporate priorities. But at the same time the signatures, often coupled with a link provided a rare peak into a vibrant creative scene spanning the Middle East and North Africa. It was a very rare moment when Israelis and Arabs were united by their shared history, and by their exclusion from "the future".

Language is the number one concern of the Arab Techies <*arabtechies.net*>, a geek network spread all over the Arab world. To a western eye their online presence seem very foreign, with the text on their site being mostly in Arabic with just a bit of English and French and with some pictures portraying mostly young people some wearing Hijab. When translated though, the actual content of the site echoes the exact ideological line of the Free Culture movement. Inspiration from Lawrence Lessig's calls for a more nuanced Creative Commons licensing regime, an overarching excitement from Free Software and a religious commitment to sing its praises, and a general optimism for how information networks can change the world. The Arab Techies are concerned about Arabic localization, they develop Open Source code libraries to address the bi-direction and translation issues. They are concerned with Arabic typography and how the highly calligraphic letters render to the screen and they share best practices for right-to-left minded design.

> No one can deny the scale of Internet and mobile phone penetration in the Arab World. People in the region are becoming increasingly aware of the potentials offered by technology for social and political change. Artists, social workers and young intellectuals are resorting to information and communication venues in order to disseminate their work, gain wider reception and create more interaction. Despite the emergence of such highly connected communities of citizen journalists, cyber artists and digital activists, the techies who provide support and infrastructure to these communities, are still working in isolation, not really benefiting from this regional networking.

> While their social role is not always recognized by their communities and sometimes even by the techies themselves, they play a pivotal role, they are builders of communities, facilitators of communication between communities, they offer support, hand holding and transfer of skills and knowledge and they are transforming into gatekeepers to an increasing diversity of voices and information.

> —*The Arab Techies Gathering* <*arabtechies.net/node/5*>

Software localization is not the only agenda of the Arab Techies. Under governments that suppress free speech, freedom of assembly and rights of self-determination the Free Culture ideology takes a very different tone. Since the censored mainstream media is not over-saturated with political debate self expression is rare and powerful. Bloggers in the Arab world have revived political debate, for that they have been arrested and tortured. Egyptians have successfully used Facebook for mass mobilization after for decades any sign decent was immediately crushed by the secret service. Knowing they are constantly followed activists use Twitter for voluntarily publishing their location making sure they cannot be "disappeared" by the government.

Arab Techies Women Gathering, Beirut May 2010

Free Culture as a Gateway Drug to Civic Engagement

But the profiles of these political activists today is different. They are not necessarily the Communist ideologues or Islamic Brotherhood hardliners. The Arab Techies like many other socially minded geeks, are first geeks and only then socially minded. Networking technologies have led them into communication and organizing. In a very apolitical way technology declares: information wants to be free. Geeks recognize that as logical, much before they see it as political. They look at old and restrictive systems and realize they cannot sustain themselves. They are radicalized by Free Culture. While in the west this would manifest as a polite call for intellectual property reform, under dictatorship this sentiment is a political time-bomb.

Free Culture in cultures that are not free is dangerous, both for those fighting for it and for those fighting against it. The free sharing of non-rivalry information goods in the west means no actual sacrifice is involved in these acts, and hence their commitment and sincerity may be questioned. Can we really say the same about those risking their lives fighting under the exact same slogans?

Free Culture and its often algorithmic logic is serving as a gateway drug to civic engagement. While in the west a lot of this engagement has already been subsumed by economic and governmental institutions, these dynamics have been working differently in Arab world. Some geeks are using the economic context of information technologies as a way of protecting themselves against from prosecution. When Tunisian free speech activists discovered a huge data surveillance conducted by their government they could not argue against it as a danger to democracy. In a country with only one party and a dubious electoral system the word democracy does not hold much water. Instead they chose to raise the concerns of the French business community raising their concerns that the Tunisian government is compromising their confidentiality when doing business in Tunisia.

Are the Arab Techies a viable example for Frosts notion of networked solidarity? While Free Culture means something else in Arabic, can its algorithmic logic transcend into political power in vastly dis-empowered civic societies? How will the social inclusion agenda jive with the rigid meritocratic rules of Free Software? And finally, when will we see these vibrant communities go beyond the translation of western ideals and develop a new local lexicon for networked collaboration. When that happens it would be up for the west to do the translation.

Epilogue

38. Anatomy of the First Book Sprint

One programmer and six writers in a room

The first creation of this book featured 6 to 8 writers in one space working through a networked software (Booki) which was simultaneously being built. Hence the architecture for the collaboration and the content produced by it were being produced at the same time.

It is difficult to over-state how difficult this could potentially be for all involved. It would be like living in a house, trying to sleep, get the kids off to school, have quiet conversations with your partner while all the time there are builders moving around you putting up walls and nailing down the floorboards under your feet. Not easy for all parties.

Working on adding new features and debugging code live while the people wanting to use it and are in the same room using it, is a pretty extreme environment for a programmer to work in.

Thankfully, we survived this particular cross-discipline collaboration between programmer and writers because the attitudes of all those involved enabled this to be a relatively stress-free environment. The generosity of spirit exhibited by all collaborators meant that this situation was not only tolerable but acceptable and even mostly fun!

As a result we have not only a book, but a vastly improved alpha version of the Booki collaborative authoring platform. And more importantly; a method.

The Calendar

Day one consisted of presentations and discussions.

During this first day we relied heavily on traditional 'unconference' technologies—namely colored sticky notes. With reference to unconferences we always need to tip the hat to Allen Gunn and Aspiration for their inspirational execution of this format. We took many ideas from Aspiration's Unconferences during the process of this sprint and we also brought much of what had been learned from previous Book Sprints to the table.

First, before the introductions, we each wrote as many notes as we could about what we thought this book was going to be about. The list consists of the following:

- When Collaboration Breaks.
- Collaboration (super) Models.
- Plausible near and long term development of collaboration tech, methods, etc. Social impact of the same. How social impact can be made positive. Dangers to look out for.
- Licenses cannot go two ways.
- Incriminating Collaborations.
- In the future much of what is valuable will be made by communities. What type of thing will they be? What rules will they have for participation? What can the social political consequences be?
- Sharing vs Collaboration.
- How to reconstruct and reassemble publishing?
- Collaboration and its relationship to FLOSS and GIT communities.
- What is collaboration? How does it differ from cooperation?
- What is the role of ego in collaboration?
- Attribution can kill collaboration as attribution = ownership.
- Sublimation of authorship and ego.
- Models of collaboration. Historical framework of collaboration. Influence of technology enabling collaboration.
- Successful free culture economic models.

Then each presented who they were and their ideas and projects as they are related to free culture, free software, and collaboration. The process was open to discussion and everyone was encouraged to write as many points, questions, statements, on sticky notes and put them on the wall. During this first day we wrote about 100 sticky notes with short statements like:

- "Art vs Collaboration"
- "Free Culture does not require maintenance"
- "Transparent premises"
- "Autonomy: better term than free/open?"
- "Centralized silos vs community"
- "Free Culture posturing"

...and other cryptic references to the thoughts of the day. We stuck these notes on a wall and after all of the presentations (and dinner) we grouped them under titles that seemed to act as appropriate meta tags. We then drew from these groups the 6 major themes. We finished at midnight.

Day two—10.00 kick off and we simply each chose a sticky note from one of the major themes and started writing. It was important for us to just 'get in the flow' and hence we wrote for the rest of the day until dinner. Then we went to the Turkish markets for burek, coffee and fresh Pomegranates.

The rest of the evening we re-aligned the index, smoothed it out, and identified a more linear structure. We finished up at about 23.00.

Day three—At 10.00 we started with a brief recap of the new index structure and then we also welcomed two new collaborators in the real-space: Mirko Lindner and Michelle Thorne. Later in the day, when Booki had been debugged a lot by Aco, we welcomed our first remote collaborator, Sophie Kampfrath. Then we wrote, and wrote a bit more. At the end of the day we restructured the first two sections, did a word count (17,000 words) and made sushi.

After sushi we argued about attribution and almost finished the first two sections. Closing time around midnight.

Day four—A late start (11.00) and we are also joined by Ela Kagel, one of the curators from Transmediale. Ela presented about herself and Transmediale and then we discussed possible ways Ela could contribute and we also discussed the larger structure of the book. Later Sophie joined us in real space to help edit and also Jon Cohrs came at dinner time to see how he could contribute. Word count at sleep time (22.00): 27,000.

Day five—The last day. We arrived at 10.00 and discussed the structure. Andrea Goetzke and Jon Cohrs joined us. We identified areas to be addressed, slightly altered the order of chapters, addressed the (now non-existent) processes section, and forged ahead. We finished 2200 on the button. Objavi, the publishing engine for Booki, generated a book-formatted PDF in 2 minutes. Done. Word count ~33,000.

Original Collaborators

The starting 7 included:

Mushon Zer-Aviv is a designer, an educator and a media activist from Tel-Aviv, based in NY. His work explores media in public space and the public space in media. In his creative research he focuses on the perception of territory and borders and the way they are shaped through politics, culture, networks and the World Wide Web. He is the co-founder of Shual.com—a foxy design studio; ShiftSpace.org—an open source layer above any website; YouAreNotHere.org—a dislocative tourism agency; Kriegspiel—a computer game based on Guy Debord's Game of War; and the Tel Aviv node of the Upgrade international network. Mushon is an honorary resident at Eyebeam—an art and technology center in New York. He teaches new media research at NYU and open source design at Parsons the New School of Design. He can be found at Mushon.com

Mike Linksvayer is Vice President at Creative Commons, where he started as CTO in 2003. Previously he co-founded Bitzi, an early open data/open content/mass collaboration service, and worked as a web developer and software engineer. In 1993 he published one of the first interviews with Linus Torvalds, creator of Linux. He is a co-founder and currently active in Autonomo.us, which investigates and works to further the role of free software, culture, and data in an era of software-as-a-service and cloud computing. His chapter on "Free Culture in Relation to Software Freedom" was published in FREE BEER, a book written by speakers at FSCONS 2008. Linksvayer holds a degree from the University of Illinois at Urbana-Champaign in economics, a field which continues to strongly inform his approach. He lives in Oakland, California.

Michael Mandiberg is known for selling all of his possessions online on Shop Mandiberg, making perfect copies of copies on AfterSherrieLevine.com, and creating Firefox plugins that highlight the real environmental costs of a global economy on TheRealCosts.com. His current projects include the co-authored groundbreaking Creative Commons licensed textbook "Digital Foundations: an Intro to Media Design" that teaches Bauhaus visual principles through design software; HowMuchItCosts.us, a car direction site that incorporates the financial and carbon cost of driving; and Bright Bike, a retro-reflective bicycle praised by treehugger.com as "obnoxiously bright." He is a Senior Fellow at Eyebeam, and an Assistant Professor at the College of Staten Island/CUNY. He lives in, and rides his bicycle around, Brooklyn. His work lives at Mandiberg.com.

Marta Peirano writes about culture, science and technology for the Spanish media, encompassing newspapers, online journals and printed magazines. She is a long term contributor and founder of the online media arts journal Elástico and is the author of "La Petite Claudine", a widely read blog in the Spanish language about art, literature, free culture, pornography (and everything in between). In 2003 and 2004 she directed the Copyfight Festivals in Spain (CCCB, Santa Mónica) with her collective Elástico, a symposium and exhibition that investigated alternative models of intellectual property. Marta has given numerous lectures and workshops on free culture, digital publishing tools and journalism at festivals and universities. She recently published "El Rival de Prometeo", a book about Automatas and the engineering of the Enlightenment. She currently lives in Berlin and is working on a second book.

Alan Toner was born in Dublin and studied law in Trinity College Dublin and NYU Law School, where he was later a fellow in the Information Law Institute and the Engelberg Center on Law and Innovation. His research is focused on the countervailing impact of peer processes and information enclosure on cultural production and social life. In 2003 he worked on the grassroots campaign 'We Seize!' challenging the UN World Summit on the Information Society; he has participated extensively in grassroots media and information freedom movements. Since 2006 he has also worked in documentary film, including co-writing and co-producing "Steal This Film 2" (2007). In 2008 he co-created the archival site http://footage.stealthisfilm.com/. Currently he's writing a book on the history of economic and technological control in the film industry. Sometimes he can be found near Alexanderplatz, and at http://knowfuture.wordpress.com/.

Aleksandar Erkalovic is renown internationally in the new media arts and activist circles for the software he has developed. He used to work in Multimedia institute in Croatia, where he was the lead developer of a popular NGO web publishing system (TamTam). Aleksander has a broad spectrum of programming experience having worked on many projects from multiplayer games, library software, financial applications, artistic projects, and web site analysis applications, to building systems for managing domain registration. Aleksander was for a long time the sole programmer for FLOSS Manuals and is now leading the development (together with Adam Hyde and Douglas Bagnall) of a new GPL-licensed type of collaborative authoring and publishing platform called 'Booki'. Aleksander's new media artistic collaborations have won many awards, as well as being extensively exhibited internationally. Aleksander also organises creative and educative workshops directed to young people, experts, and amateurs that are interested in the software he has developed and free software in general. He is currently also employed by Informix in Zagreb, Croatia.

Adam Hyde <adam[at]flossmanuals.net> was for many years a digital artist primarily exploring digital-analog hybrid broadcast systems. These projects included The Frequency Clock, Polar Radio, Radio-Astronomy, net.congestion, re:mote, Free Radio Linux, Wifio, Paper Cup Telephone Network, Mobicasting, Silent TV and others. Many of these projects have won awards and have been widely exhibited internationally. Since returning from a residency in Antartica in 2007 Adam founded FLOSS Manuals and has been focused on increasing the quantity and quality of free documentation about free software through FLOSS Manuals, exploring emerging methodologies for collaborative book production (Book Sprints), and developing Booki with Aleksander and Douglas. Adam has facilitated over 16 Book Sprints, is also the co-founder (with Eric Kluitenberg) of the forthcoming Electrosmog Festival for Sustainable Immobility and facilitator of the forthcoming Arctic Perspectives technology cahier.

The cover design is by Laleh Torabi. Laleh is a designer and illustrator based in Berlin and has been the designer for Transmediale for several years. Her website is <www.spookymountains.com>. Her latest book "Die Freiheit der Krokodile" (The Freedom of the Crocodiles) has just been released by Merve Verlag, Berlin.

Those that joined later include:

Ela Kagel—an independent cultural producer and curator in Berlin. She is curator of Public Art Lab, initiator of Upgrade! Berlin, co-initiator of Mobile Studios and program curator of transmediale10.

Michelle Thorne—International Project Manager at Creative Commons, coordinating over fifty jurisdictions worldwide to localize and promote the Creative Commons licensing suite worldwide. Michelle co-organizes the Berlin salon series OpenEverything Fokus and also the network and festival atoms&bits. She holds a B.A. in Critical Social Thought and German Studies from Mount Holyoke College and is based in Berlin, Germany.

Mirko Lindner—an Open Everything advocate, active in FLOSS, Free Culture as well as Copyleft Hardware. His involvements range from Creative Commons Sweden over FSCONS to paroli on the Neo Freerunner. His main project right now is Sharism at Work. His areas include communication, design, planning as well as small-scale development surrounding the Ben NanoNote and the company's infrastructure. Mirko is a founding member of Sharism at Work.

Sophie Kampfrath—a Berlin based student of German literature, linguistics and philosophy. Being interested in new ways of working evoked by web technologies, she joined the atoms&bits network. Atoms&bits is about the impact of virtual and net developments on, and interleaves with, the physical world. Another aspect of this is the work on and with the Hallenprojekt, a platform and network bringing together co-working spaces and people.

Jon Cohrs—a recording engineer and visual/sound artist who lives in Brooklyn, NY. Through residencies, installations, and performances at I-Park, Banff New Media Institute, Futuresonic, and Eyebeam, his work has focused on exploring technology and its connection with wilderness through his documentary "The Door to Red Hook: Backpacking through Brooklyn", his website ANewF*ckingWilderness.com, and the 2009 Futuresonic Art Award winner, the Urban Prospector. Most recently, he's been an artist in residence at the Eyebeam Atelier working on 'OMG I'm on .TV'. This is an analog Pirate TV station in New York City that fills the void left behind after the digital transition, addressing the evolution of media, fan based culture, copyrights, and discussions on bandwidth allocation. OMG TV was used as a reference in a Supreme Court amicus brief on creativity and copyright.

Andrea Goetzke—Berlin-based curator, consultant and organizer of events and projects, and part of newthinking communications. She has engaged in several projects in the area of open source approaches and digital culture, like the Openeverything event series or the all2gethernow camp, a participatory event on new strategies in music and culture.

Patrick Davison contributed the opening chapter remotely (from New York). Patrick is a digital artist living in Brooklyn, NY. As one half of group What We Know So Far he researches and presents on Internet memes, digital ephemera, modern information culture, community, love, and time travel. He works with Eyebeam Senior Fellow Michael Mandiberg to research, create, and document work, and collaborated with FLOSS Manuals during their participation in Wintercamp 2009.

Jonah Bossewitch contributed the Multiplicity and Social Coding Chapter via email. Jonah is a doctoral candidate in Communications at Columbia's School of Journalism. He also works full-time as a technical architect for Columbia's Center for New Media Teaching and Learning (CCNMTL). He is investigating the politics of memory, surveillance, and transparency and their intersection with corruption in the pharmaceutical industry. Jonah has over a decade of experience as a professional free software developer and a vocal advocate for free culture, mad pride, and social justice. He completed an MA in Communication and Education at Teachers College ('07) and graduated from Princeton University ('97) with a BA Cum Laude in Philosophy and certificates in Computer Science and Cognitive Studies. He blogs at alchemicalmusings.org.

39. 2 Words vs. 33,000

Over the course of the second book sprint we often paused to reflect on the fact that editing and altering an existing book (one originally written five months prior by a mostly different group of people) is a completely different challenge than the one tackled by the original sprinters. While the first author group began with nothing but two words -Collaborative Futures-, words that could not be changed but were chosen to inspire. This second time we started with 33,000 words that we needed to read, understand, interpret, position ourselves in relationship to, edit, transform, replace, expand upon, and refine.

Coming to a book that was already written, the second group's ability to intervene in the text was clearly constrained. The book had a logic of its own, one relatively foreign to the new authors. We grappled with it, argued with it, chipped at it, and then began to add bits of ourselves. On the first day the new authors spent hours conversing with some of the original team. This continued on the second day, with collaborators challenging the original text and arguing with the new contributions.

*Glossary: **Imaginary Reader***

If this book is a conversation, then reading it could be described as entering a particular state of this exchange of thoughts and ideas. Audience might be a word, a possibility and potential to describe this reader-ship; an audience as in a performance setting where the script is rather loose and does not aim for a clear and definite ending. (It is open-ended by nature); an audience that shares a certain moment in the process from a variable distance. The actual book certainly indicates a precise moment, thereby it IS also a document, manifesting some kind of history in/of open source and counter-movements, media environments, active sites, less active sites, interpassivity (Robert Pfaller <*www.psychomedia.it/jep/number16/pfaller.htm*>), residues of thought, semi-public space; history of knowledge assemblages (writers talked about an endless stitching over...) and formations of conversations. The book as it is processed in a sprint, is a statement about and of time. The reader or audience will probably encounter the book not as a "speedy material". Imaginary Readers, Imaginary Audience.

We came to recognize, however, that the point was not to change the book so that it reflected our personal perspectives (whoever we are), but to collaborate with people who each have their own site of practice, ideology, speech, tools, agency. In service of a larger aim, none of us deleted the original text and replaced it to reflect our distinct point of view. Instead, we came to conceive of *Collaborative Futures* as a conversation. Since the text is designed to be malleable and modifiable, it aims to be an ongoing one. That said, at some point this iteration of the conversation has to stop if a book is to be generated and printed. A book can contain a documented conversation, but can it be a dynamic conversation? Or does the form we have chosen demand it become static and monolithic?

In the end, despite our differences, we agreed to contribute to the common cause, to become part of the multi-headed author. Whether that is a challenge to the book or a surrendering to it, remains unclear.

Collaborators

The June 2010 sprint introduced three more core authors who worked in person and remotely with the January 2010 team:

Sissu Tarka is an artist and researcher based in London and Iceland. She currently works independently, exploring questions of the criticality of emerging practices, ethics and economies of art. Her particular focus is on non-linearity, modes of resistance, and articulations of the democratic, active work. Past investigations include micro-projects such as her text exchange with Heath Bunting during one of his border-performances, with the resulting essay *BorderXing: Heath Bunting, Sissu Tarka*, Afterall Online (2009); or *an InviTe For mAking OrnAmentS*, a workshop with Merce Rodrigo Garcia, on assemblages in architecture and technologically informed environments and networks, with a contribution by Japanese architects SANAA Kazuyo Sejima + Ryue Nishizawa, Serpentine Pavilion/Café, London (2009).

kanarinka, a.k.a. Catherine D'Ignazio (kanarinka.com), is an artist and educator. Her artwork is participatory and distributed—a single project might take place online, in the street and in a gallery, and involve multiple audiences participating in different ways for different reasons. Her practice is collaborative even when she says it's not. Her artwork has; been exhibited at the ICA Boston, Eyebeam, MASSMoCA, and the Western Front among other locations. She is Co-Director of the experimental curatorial group iKatun and a founding member of the Institute for Infinitely Small Things. After spending eight years in educational technology as a java programmer & technical project manager, she now teaches at RISD's Digital Media Graduate Program. The former Director of Exhibitions at Art Interactive in

Cambridge, MA, kanarinka maintains an experimental curatorial practice through her work with iKatun in organizing the occasional exhibition, festival or screening, and, more recently, the Platform2 event series. kanarinka has a BA in International Relations from Tufts University (Summa Cum Laude, Phi Beta Kappa) and an MFA in Studio Art from Maine College of Art. She has lived and worked in Paris, Buenos Aires, and Michigan, and currently resides in Boston, MA.

Astra Taylor (www.hiddendriver.com) is a writer and documentarian born in Winnipeg, Manitoba and raised in Athens, Georgia. She was named one of the 25 New Faces to Watch in independent cinema by Filmmaker Magazine in the summer of 2006. She co-directed "The Miracle Tree," a short documentary about infant malnutrition in Senegal, and associate produced "Persons of Interest" (Sundance 2004), about the round up and detention of Muslims and Arabs in the aftermath of September 11th. Her first film, "Zizek!," screened at festivals, in theaters, and on television around the world and was broadcast on the Sundance Channel. "Examined Life," a series of excursions with contemporary thinkers, premiered at the Toronto International Film Festival in 2008 before opening theatrically. A companion book is available from The New Press. Astra has also contributed to Monthly Review, Adbusters, Salon, Alternet, The Nation, Bomb Magazine and other outlets.

40. Knock Knock

Around noon on the second day of the First Book Sprint we hear a knock on the door. Here is the set up, we're working from a hotel room in a complex called IMA Design Village, on the 5th floor of a redeveloped late 19th Century factory building with a jerky elevator and nothing to indicate where we are. All of us were in the room at the time and we were not expecting company. We opened the door and there stood a guy around our age who said he heard about the project and he wanted to contribute.

We were all amazed: the writers and the guy in the hallway. But mainly we were unprepared for this. He didn't even say his name, he just said he had some ideas about collaboration and he really wanted to contribute. That was just completely great! But while we announced that the collaboration will be later opened to remote collaboration, at that moment, in that place we were completely unprepared for more people in the room. The anonymous contributor said he had met Adam at an obscure music event in Berlin. Adam and the anonymous contributor went downstairs to the cafe to discuss how he could contribute. It was planned for him to write some material remotely and possibly join us the following day.

Glossary: **Location-Locating**

> Includes: placing, territories, deterritorialisation, context, site, cities such as Berlin and New York, places such as playgrounds, airports, mobile stations, ice-scapes. Contexts such as festivals, exhibitions, neighborhoods and conferences.

This was a unique experience of finally meeting the epic "anonymous user" in person. That faceless person that does not even have a username but is highly motivated and just wants to start contributing was standing there in-person at our doorstep. We didn't know his name, we only knew his IP address—where he physically is: he was right there! Literally browsing our "collaborative site".

And we? We were so Alpha, we were what early web people two decades ago used to call "under construction" or "in stealth mode." We didn't even have an interface for him yet. It's like he found a public yet unannounced URL for a future collaborative platform that was just not ready yet. We thought we were private, but apparently we were live. We were caught off-guard with our first anonymous visitor, very online and just eager to log in.

41. Are we interested?

The issue of subjectivity quickly turned up in the first book sprint. Mushon was about to write: "I am actually more interested in…" But since we decided to write in plural, he pulled his head out of the screen and asked: "How should I write this? Am I interested? Are WE interested?"

The following day this conflict was raised again, this time it was even more complex when Mushon wanted to refer to a personal anecdote. Both Michael and Mike have already done it in their own writing but they were able to quote themselves as they were indeed quoting previous published text. In this specific case, Mushon was recounting the grim memory from his army days that is mentioned in the chapter titled "Collaborationism". This was the first time he has ever put it in writing.

Should he write "I"? Who is "I"? We're writing in plural, as "we". Should he write "one of the authors"? That's pretty superficial, and even ridiculous. How many of "the authors" have served in the Israeli army? Should he quote himself? It doesn't really make sense, it is not like he is re-appropriating a quote from a previously published piece. He proposed to explicitly declare he is switching to first person for the sake of a personal anecdote, but that posed a stylistic problem.

It is just an anecdote, any writer will just write it as: "I remember…". Is language just not equipped for collective writing? Will more experiments like this one force a new way of elegantly switching between group and individual identities?

It seems like for now we will leave it is as is—unstated. If in 30 years or so the English language finally catches up and will come up with new linguistic tools for collective writing, feel free to edit.

 Glossary: **Vocabularies**

> For specific fields, for expressing one's belonging, for translations. Enhance communication, questions, specificities, dialogue, and debates where things can be talked about. Think this book as a vocabulary!

---

42. Sample Chat

"CollaborativeFutures" is being published.

mushon: @mike

mushon: You wrote:Other old examples that are in many ways more interesting examples of collaboration than their modern counterparts include IRC (Twitter) and Usenet (forums).

MikeLinksvayer: true

mushon: I wonder if we can switch the "more interesting" to something else

mushon: I actually think that while Twitter is informed by IRC, it is very different

mushon: ...

MikeLinksvayer: sure

mushon: I'll edit it a bit and show you, ok?

MikeLinksvayer: it does say "in many ways", not absolutely

MikeLinksvayer: but i agree with going further in that direction

MikeLinksvayer: go ahead

mushon: I do understand the context and the point you and Adam are trying to make

mushon: so it will hopefully just focus it

MikeLinksvayer: yep

INFO "CollaborativeFutures" is being published.

MikeLinksvayer: i'm not emotionally attached at all. :)

mushon: ;)

JOINED booki

INFO User adamhyde has created new chapter "Sample Chat".

INFO User MikeLinksvayer has changed status of chapter "Outsiders: thoughts on external collaboration" to "Written".

INFO User adamhyde has saved chapter "Sample Chat".

================================

*Glossary: **Discourse***

Sometimes useful to generate new meanings, ideas, images, codes, sources, open sources. Sometimes blocks immediate action, and affects.

INFO User mandiberg has saved chapter "Does Aggregation Constitute Collaboration?".

INFO User mandiberg has changed status of chapter "Does Aggregation Constitute Collaboration?" to "Editied".

INFO User AlanToner has saved chapter "Motivations for Collaboration".

INFO User adamhyde has saved chapter "About this Book".

INFO User adamhyde has saved chapter "About this Book".

JOINED AlanToner

INFO User AlanToner has saved chapter "Motivations for Collaboration".

INFO User ela has saved chapter "Collaborative Economies".

INFO User adamhyde has saved chapter "About this Book".

INFO User Marta has saved chapter "Generosity".

INFO User ela has saved chapter "Collaborative Economies".

INFO User adamhyde has saved chapter "Boundaries of Collaboration".

JOINED sophie_k

INFO User mushon has saved chapter "Anonymous".

INFO User adamhyde has saved chapter "The New York Special Edition".

INFO User mandiberg has saved chapter "Anonymous".

INFO User adamhyde has renamed chapter "The New York Special Edition" to "Attribution Imbalance".

JOINED Marta

INFO User mandiberg has saved chapter "Introduction".

JOINED Marta

INFO User adamhyde has saved chapter "Boundaries of Collaboration".

INFO User sophie_k has saved chapter "Collaborative Economies".

INFO User mandiberg has saved chapter "Anonymous".

INFO User mandiberg has saved chapter "Anonymous".

INFO User mandiberg has saved chapter "Anonymous".

INFO User adamhyde has saved chapter "How this book is written".

INFO User Marta has saved chapter "Anonymous Collaboration".

JOINED agoetzke

INFO User adamhyde has saved chapter "Boundaries of Collaboration".

INFO User mandiberg has saved chapter "Crowdfunding".

INFO User adamhyde has renamed chapter "Problematizing attribution" to "Problematizing Attribution".

JOINED booki

INFO User AlanToner has saved chapter "Setting the Future Free: Ownership, Control and Conflict".

MikeLinksvayer: Why Science 2.0? Didn't we claim in the chapter A Brief History of Collaboration that "Web 2.0 is bullshit" as the "version number" of the web as it conveys the incorrect sense that progress is not incremental and a marketing-driven message to "upgrade"? For these same reasons Science 2.0 is appropriate. In general science hasn't made effective use the web -- translating and adopting the best practices of open collaboration on the web would consitute an "upgrade" and such an upgrade should be encouraged rhetorically. This is largely the case due to science's current setting in giant, slow to change institutions -- "big science". But, institutions, when they do change, can force broad change, quickly, as a matter of policy. Another reason the "upgrade" connotation is appropriate.

INFO User MikeLinksvayer has saved chapter "Science 2.0".

INFO User AlanToner has saved chapter "Setting the Future Free: Ownership, Control and Conflict".

INFO User adamhyde has saved chapter "Science 2.0".

JOINED AlanToner

INFO User Marta has saved chapter "Death is not the end".

INFO User Marta has saved chapter "Death is not the end".

INFO User AlanToner has saved chapter "Setting the Future Free: Ownership, Control and Conflict".

INFO User mandiberg has saved chapter "Other People's Computers".

INFO User mandiberg has changed status of chapter "Other People's Computers" to "Editied".

INFO User mushon has saved chapter "Solidarity".

INFO User adamhyde has saved chapter "Death is not the end".

INFO "CollaborativeFutures" is being published.

INFO User adamhyde has saved chapter "Looking in from the outside".

JOINED sophie_k

INFO User Marta has saved chapter "Death is not the end".

JOINED PatrickDavison

JOINED booki

INFO "CollaborativeFutures" is being published.

INFO User Marta has saved chapter "Death is not the end".

INFO User adamhyde has saved chapter "Setting the Future Free: Ownership, Control and Conflict".

INFO User Marta has saved chapter "Death is not the end".

INFO User Marta has saved chapter "Death is not the end".

INFO User mandiberg has saved chapter "Collaborative Economies".

INFO User mushon has saved chapter "Solidarity".

INFO User adamhyde has saved chapter "Beyond Education".

INFO User sophie_k has saved chapter "Problematizing Attribution".

INFO User AlanToner has saved chapter "The Freedom to Merge, The Freedom to Fork".

INFO User AlanToner has saved chapter "The Freedom to Merge, The Freedom to Fork".

mandiberg: yo patrick!

PatrickDavison: yo

PatrickDavison: what up?

INFO User MikeLinksvayer has saved chapter "A Brief History of Collaboration"

INFO User Marta has saved chapter "Crowdfunding".

mandiberg: i rewrote b/c we restructured the book

mandiberg: it became about using it as an intro to the themes

PatrickDavison: yeah the rewrite is great.

mandiberg: we are trying to lock the book

mandiberg: just fact check me

mandiberg: make sure i've got my names and urls and terms right

mandiberg: kk?

mandiberg: we only have 1hr to finish the whole thing

PatrickDavison: OH! word.

PatrickDavison: yeah I'm polishing up.

INFO User adamhyde has saved chapter "Sample Chat".

================================

adam hyde: hwosit all going

kanarinka: hey adam - we were just talking about the "This BOok is Useless" section

adam hyde: goodo

kanarinka: and Mushon was saying that maybe it was being perceived as

kanarinka: too critical of the previous project

kanarinka: (the title, not the section)

kanarinka: but I just wanted to explain a bit

adam hyde: ok

kanarinka: that it's both a way of raising a question and being provocative at the same time

adam hyde: i think the title is ok, but the point isnt strongly made yet in the text i think

kanarinka: the main point is really about relationality underlying everything we already do

kanarinka: have you looked recently? I just updated it this AM

adam hyde: ok, lemme look again

kanarinka: it was a little disorganizing bc i wasn't sure what it meant to be a chapter vs a section etc

adam hyde: ok

kanarinka: so it's an open question but it comes from a place of deep commitment to collaboration and to the project

kanarinka: and i'm not REALLY asserting that the book is useless

kanarinka: more like a rhetorical device

adam hyde: yeah

adam hyde: but i think the point itself needs more weight

adam hyde: its a bit of an easy shot at the moment, needs more substance and nuance...at the moment, i see what it is trying to say but it doesnt capture my imagination like i think it should

kanarinka: ok - tell me more

adam hyde: the book doesnt assert that we never collaborate. and i think this is what this chapter is implying

adam hyde: tit doesnt seem to grasp the fact that we are talking about particular kinds of collaboration

adam hyde: possibly limited to the free culture digital media zone

adam hyde: and that is a limit for sure

kanarinka: well the idea of making a book about collaboration imagines that there is something such as not collaborating

adam hyde: does it?

kanarinka: yes

adam hyde: im not so sure

adam hyde: i think u can talk about strong and weak collaboration

adam hyde: and collaboration in specific sphres

adam hyde: the original text does not say that people exist that never collaborate in all contexts

adam hyde: it talks about specific contexts

kanarinka: specific spheres as in free culture digital media

kanarinka: you mean?

adam hyde: yes

adam hyde: prety much

adam hyde: i think u need to recognise that a little

adam hyde: else the bookl, in my opinion, goes everywhere at once

adam hyde: which isnt so interesting i think, at least not to me

kanarinka: i think it's important to acknowledge a wider framework and context of collaboration

kanarinka: as well as a lot of the underlying assumptions that the book makes

adam hyde: yes, but wiht reference to the context that is being discussed

kanarinka: about individuals, agency, and so on

kanarinka: i am raising questions as to whether we can even speak of individuals collaborating with each other

kanarinka: since it is something we always do in all cases at all times

kanarinka: and so what is there to say

kanarinka: however

kanarinka: that said

kanarinka: i do think there is something to be said

kanarinka: or i wouldn't be here!

===============================

astrataylor: do you think it would be possible to make a truly collaborative film -- an open collaboration where anyone could participate on any level -- that was not "creatively poor" but really raised the artistic bar? i'd like to think so, but it would have to be a very special group i suspect....and a group that maybe agreed on aesthetics and issues of form and was therefor a bit homogeneous

AlanToner: certainly in terms of the downloading of mainstream hollywood film where unauthorised sharing simply contributes to the reproductioin of that model's hegemony

AlanToner: ah

AlanToner: well that's what i wanted to get around regarding the STF archive

Mushon: stop writing each other, start writing the book!

Mushon: ;)

AlanToner: I think that you can do something really collaborative on the 'origination' of ilm, and rckon that it is actually imperative in terms of where documentary budget costs go

AlanToner: hands off mushon! this could go somehwere towards the futures section, and we are writing1

AlanToner: !

Mushon: :x

================================

INFO User mushon has renamed chapter "Sample Chat" to "Chat Samples".

43. Looking in from the outside

This book project already had a lot of supporters and willing contributors long before the first drop of digital ink was shed. Through a short-term outreach effort, a small but passionate group of Berliners and people abroad were interested and excited to contribute to "the free culture book sprint". The process as well as the topic triggered a great response online. Already a week before the book sprint began, about ten people met in a local cafe to discuss how they could contribute. Adam explained the process and philosophy behind FLOSS Manuals and this project in particular, and we also spoke to representatives from the Transmediale festival. In general, there was an enthusiasm and buzz about writing a collaborative book.

Following the meeting, Adam gave some thought to the role of "external contributors". It was clear from the book's title that it would not be as straightforward as a software manual. "Collaborative Futures" is a complex topic, which could take on many manifestations and directions. It's an experimental topic for a book sprint—who knew how it would go? To err on the safe side, it was decided that external visits to the book sprint location would be limited to those who could commit to at least one full day. The table of contents would be posted online, but only after the core group of authors had produced it on the first day.

Beginning on the second day, it was announced that external contributors could expect the table of contents, and from that they could find a section or topic that interested them and write. However, once Tuesday came around and the index was circulated, it was difficult for anyone who wasn't in the room to understand what was going to be written, how it was structured, and what the skeleton meant. There were no guidelines or notes to follow that would have really helped outsiders find a voice in the project or a meaningful place to contribute. Still, a number of people were still interested in the project and wanted to help. We didn't know where and how to direct that energy.

When we (Mirko and Michelle) arrived at the location on the third day, we were warmly welcomed and introduced to the group. Everyone was engaged at their computers, ready to write, but friendly and open. After a round of introductions and a brief overview of the table of contents, we were asked how we'd like to contribute. We selected areas of interest which seemed complementary and relevant (collaboration between companies & community and co-working), and got to work.

After a few hours of intense writing and reading, we were finding it difficult to frame and articulate our sections in a meaningful way. We were realizing that the group, within the span of three days, had developed its own language. They had a streamlined plan for their writing, and they understood each others' arguments and tasks. It is incredibly impressive that such a diverse group had converged on that level of consensus in such a short time.

Nevertheless, it proved challenging to tie our writing into the group's larger narrative. We were not equipped with the language nor knowledge about fundamental decisions they already made. It was hard to build upon their themes and connect ideas. In group conversations and one-on-one, our suggestions were welcomed and heard, but there was still a gap in the modes of writing.

It's important to emphasize that this is not a pariah problem. There was just already a very intense and productive atmosphere of collaboration. Ideas and suggestions were flying, people were working solidly on their sections, and time was of the essence. It seemed as if the information collection phase was complete, and now it was time to write.

Was the sprint a victim of the mythical man month? Did adding new people slow the process? We hope not. But we still found it hard to evaluate and properly place our contributions, even if we spent an intense day with them reading, writing, and discussing.

Already on day three, the group had produced an ambitious outline and an immense amount of text. Our pitstop visit, already late in the process, meant that our contributions would not be fundamental. Instead, we could merely suggest, fine-tune, flesh out, etc., but the momentum was so great that there was nothing "significant" left for the new kids. This raised the question for us: what could be our real contributions?

The question of attribution and valuing contributions was a theme brought up heatedly by the group later that evening. In a project like a book sprint, with six core authors and some external contributions, how can you scale recognition? What's the best policy for doing so? It is an open question, and one we think FLOSS Manuals handles well. But nevertheless it is good food for thought.

So, as the group plows on for the next two days, we ask ourselves: how can this process be improved? How can the energy and knowledge of external contributors, people not within the core author group, be put to good use? We've brainstormed about some options, some of which may not have been fitting for this particular sprint, but may nevertheless be helpful for future projects.

- The hardest thing for outsiders is understanding the "language" of the core group. Taking notes, publishing more supplementary material, and clarifying the goal and scope of the book would make it easier for external people get a handle on the project.
- Tasks and needs should be clearly articulated by the core group. Do you need editing help? An expert on a certain protocol? Research? Explain what you need, and there may be the expertise and skills outside to help.
- Writing could be scheduled to include a comment period for external contributors. For example, after the second day, a certain section could be submitted to the public, discussed outside, and then revisited on the next day. The fresh perspective could be useful.
- Another idea from someone who only joined on day 5 and very much agrees to this chapter (Andrea): While it is difficult for an outsider to fully get into the flow, tone and stream of thought of an intensely collaborating group, there could be an annotation section for each chapter, where outsiders can contribute additional examples and thoughts, which then the core group can consider for discussion and editing.

All in all, we really had fun and experienced a book sprint firsthand. We wish we could have helped more, but we appreciate the process and have learned from it. We are very grateful for the chance to join this great group and meet the people behind this book. We also are thankful for all the other external contributors who helped the project. This is an evolving process, and we are happy to have been a part of it!

Appendices

44. Things we ended up not including

Some of these were out of scope, some we didn't have time to include. All could become in scope for a future edition, to be evaluated by future collaborators.

- Crowdsourcing & Mechanical Turk
- Internal collaboration in for-profit businesses
- Relative maintenance efforts of collaborative and free culture projects
- Interns
- FLOSS zealotry and License fascism and Free Culture as an atheistic faith
- Free Culture posturing, and not walking the talk
- Scaling collaborations
- Tolerance of errors
- The pain of confronting ideologies
- How to collaborate with people you don't agree with
- Ego
- Resilient communities
- Previous sprint-like collaborative writing projects (*Unnatural Acts*)
- Freedom to vs. freedom from (bottom-up vs. top-down) and the seamless transition between the two (forks and merges)

45. Write this Book

The greatest irony of the collaboration that produced the 1st edition of Collaborative Futures was its partial failure to incorporate collaborators beyond the core group that spent a week in Berlin working face to face. As recounted in the 1st edition's epilogue chapters <*en.flossmanuals.net/CollaborativeFutures/KnockKnock*> written by collaborators in Berlin who did not start with the core group did not "fit" (with the valuable exception of their recounting of not fitting!) and a walk-in collaborator could not be accommodated. It proved impossible to open up the real-time collaboration to potential remote collaborators. However, some additional collaborators in Berlin helped with copy editing and one chapter was contemporaneously written by a friend (Bossewitch) of one of the core collaborators (Zer-Aviv), shepherded by that collaborator.

The 2nd edition sprint mandated increased temporally and geographically distributed collaboration as it was built on the 1st edition and was structured as a face to face sprint in New York with remote contributors from the 1st edition. However, no drop-in contributions were realized.

Thus, herewith are some possible practices for future sprint teams and others, whether coordinated with sprints or as the book is discovered and someone is inspired to make a substantial contribution. These practices aren't gospel; they hopefully aren't mostly wrong and are definitely subject to revision.

For new collaborators, any venue, any time

Read the previous edition. This is the best way to ensure your work will complement the existing text—whether your work is to be complimentary, critical, or expanding.

For all collaborators, any venue, any time

If you're not sure how to contribute and perhaps not sure who to ask what is needed, here are some valuable activities that require little coordination:

- Copy-edit existing text
- Annotate existing text, e.g., by adding references where needed or by finding images that illustrate the text
- Write a completely new chapter; actual and speculative case studies in particular can be independent, but having read the book, can be tied to existing themes

Whether you have a clear idea for your contribution or not, keep good collaboration practices in mind (if you notice that an important practice isn't discussed in the book, there's your chapter to write). Assume good faith.

Sprinters, with additional ideas for multi-location sprinters

Possibly the most challenging part of the 1st sprint was the start, in which the core group, starting with two words, decided what to write about and generally mind-melded. The necessary success here (it is easy to imagine failure) probably contributed to the difficulty of adding collaborators.

Imagine an nth edition with sprints in São Paulo and Nairobi aim to substantially restructure the book or pursue a divergent theme—as opposed to diving into the valuable but low-coordination work mentioned above—it would be good for the two teams to agree in broad strokes to the path forward and be able to communicate that path to each other—and to remote collaborators. Some ideas:

- Discuss direction of the sprint prior to the sprint days. The first sprint did no pre-work in part to prove a point—a non-manual could be successfully written in a week starting from only a two word theme. There is no reason for subsequent sprints or other forms of contribution to avoid pre-work, excepting lack of time.
- There will be a few (to many if Nairobi sprinters work late) hours of overlap each day between the São Paulo and Nairobi sprints. On the first day, it may be useful for all sprinters to give a few minute self-introduction—this was valuable for information and rapport gained in the first sprint. Throughout, it may prove valuable to have voice, preferably enhanced with video, communication on-tap for higher bandwidth cross-sprint discussion.
- When a team finishes for an evening, they should leave brief notes about changes and discoveries made, for maximum continuity during periods in which only one team is working.

Remote sprinters

Remote sprinters may wish to stick with the low-coordination contributions listed above—success along these lines would be extremely valuable. If the face-to-face teams are establish super communications, a side effect could be increased ability of remote collaborators to contribute even where higher coordination is required.

For ongoing collaboration

Communication among collaborators beyond the Booki editor is potentially key in the scenarios above, perhaps even more so for ongoing collaboration. Some mechanisms:

- Get in touch with other collaborators by posting to the FLOSS Manuals mailing list <*lists.flossmanuals.net/listinfo.cgi/discuss-flossmanuals.net*>
- Other FLOSS Manuals projects have established practices that will be useful for ongoing work on this book—see FLOSS Manuals maintainer documentation <*en.flossmanuals.net/FLOSSManuals/Maintainers*>

- If you want to take the book in an entirely new direction, you may do so under the terms of the book's CC BY-SA license. You could fork on Booki.cc, on your own instance of Booki, or by importing into a technical environment of your choosing. It would be nice to tell us about your fork or your consideration of one, but you don't have to; we'll find out when you link to the existing book's web page, per the attribution term of CC BY-SA!

Source materials

- Book "source" may be accessed (and edited) at <*www.booki.cc/collaborativefutures/edit/*>
- Mailing list signup, cover art and stylesheets may be found at <*www.collaborative-futures.org*>

Good luck!

46. Credits

Anonymous
© Collaborative Futures Book Sprint team 2010
Modifications:
PatrickDavison - 2010
mandiberg - 2010
mushon - Mushon Zer-Aviv 2010
sophie_k - 2010
adamhyde - adam hyde 2010
AlanToner - 2010
Marta - 2010
Equisetum - 2010
kanarinka - kanarinka 2010
MikeLinksvayer - Mike Linksvayer 2010

How This Book is Written
© Collaborative Futures Book Sprint team 2010
Modifications:
adamhyde - adam hyde 2010
vegyraupe - 2010
sophie_k - 2010
agoetzke - 2010
Marta - 2010
Equisetum - 2010
kanarinka - kanarinka 2010
MikeLinksvayer - Mike Linksvayer 2010
mushon - Mushon Zer-Aviv 2010
mandiberg - 2010
AlanTon - Alan Toner 2010

A Brief History of Collaboration
© Collaborative Futures Book Sprint team 2010
Modifications:
mandiberg - 2010
adamhyde - adam hyde 2010
MikeLinksvayer - Mike Linksvayer 2010
mushon - Mushon Zer-Aviv 2010
thornet - 2010
AlanToner - 2010
sophie_k - 2010

mandiberg - 2010
sophie_k - 2010
MikeLinksvayer - Mike Linksvayer 2010

Criteria for Collaboration
© Collaborative Futures Book Sprint team 2010
Modifications:
AlanToner - 2010
adamhyde - adam hyde 2010
mushon - Mushon Zer-Aviv 2010
MikeLinksvayer - Mike Linksvayer 2010
sophie_k - 2010
Marta - 2010
AlanTon - Alan Toner 2010
mandiberg - 2010

Continuum Sets for Collaboration
© Collaborative Futures Book Sprint team 2010
Modifications:
adamhyde - adam hyde 2010
AlanToner - 2010
MikeLinksvayer - Mike Linksvayer 2010
AlanTon - Alan Toner 2010
marta - Marta Peirano 2010
mandiberg - 2010
mushon - Mushon Zer-Aviv 2010

Non-Human Collaboration
© Collaborative Futures Book Sprint team 2010
Modifications:
kanarinka - kanarinka 2010
collaborationist - astra taylor 2010
adamhyde - adam hyde 2010
MikeLinksvayer - Mike Linksvayer 2010
sissu - sissu tarka 2010
mushon - Mushon Zer-Aviv 2010
mandiberg - 2010
marta - Marta Peirano 2010

Boundaries of Collaboration
© Collaborative Futures Book Sprint team 2010
Modifications:
adamhyde - adam hyde 2010
sophie_k - 2010
Marta - 2010
AlanToner - 2010
MikeLinksvayer - Mike Linksvayer 2010

MikeLinksvayer - Mike Linksvayer 2010
sissu - sissu tarka 2010
mushon - Mushon Zer-Aviv 2010
marta - Marta Peirano 2010

Are We Interested?
© Collaborative Futures Book Sprint team 2010
Modifications:
mushon - Mushon Zer-Aviv 2010
adamhyde - adam hyde 2010
mandiberg - 2010
sophie_k - 2010
AlanToner - 2010
MikeLinksvayer - Mike Linksvayer 2010
sissu - sissu tarka 2010
marta - Marta Peirano 2010

Chat Samples
© Collaborative Futures Book Sprint team 2010
Modifications:
adamhyde - adam hyde 2010
mushon - Mushon Zer-Aviv 2010
kanarinka - kanarinka 2010
sissu - sissu tarka 2010
marta - Marta Peirano 2010

Looking In From the Outside
© Collaborative Futures Book Sprint team 2010
Modifications:
thornet - 2010
adamhyde - adam hyde 2010
agoetzke - 2010
mandiberg - 2010
AlanToner - 2010
MikeLinksvayer - Mike Linksvayer 2010
marta - Marta Peirano 2010

Things We Ended Up Not Including
© Collaborative Futures Book Sprint team 2010
Modifications:
mandiberg - 2010
adamhyde - adam hyde 2010
mushon - Mushon Zer-Aviv 2010
AlanToner - 2010
MikeLinksvayer - Mike Linksvayer 2010
marta - Marta Peirano 2010

Thanks for reading!

Visit http://flossmanuals.net to make corrections or to find more manuals.

ISBN 9780984475018